Jing Yang

Sprites, Lightning and Thunderstorms

Jing Yang

Sprites, Lightning and Thunderstorms

Case study of sprites-producing and non-sprites producing summer thunderstorms

LAP LAMBERT Academic Publishing

Impressum / Imprint
Bibliografische Information der Deutschen Nationalbibliothek: Die Deutsche Nationalbibliothek verzeichnet diese Publikation in der Deutschen Nationalbibliografie; detaillierte bibliografische Daten sind im Internet über http://dnb.d-nb.de abrufbar.
Alle in diesem Buch genannten Marken und Produktnamen unterliegen warenzeichen-, marken- oder patentrechtlichem Schutz bzw. sind Warenzeichen oder eingetragene Warenzeichen der jeweiligen Inhaber. Die Wiedergabe von Marken, Produktnamen, Gebrauchsnamen, Handelsnamen, Warenbezeichnungen u.s.w. in diesem Werk berechtigt auch ohne besondere Kennzeichnung nicht zu der Annahme, dass solche Namen im Sinne der Warenzeichen- und Markenschutzgesetzgebung als frei zu betrachten wären und daher von jedermann benutzt werden dürften.

Bibliographic information published by the Deutsche Nationalbibliothek: The Deutsche Nationalbibliothek lists this publication in the Deutsche Nationalbibliografie; detailed bibliographic data are available in the Internet at http://dnb.d-nb.de.
Any brand names and product names mentioned in this book are subject to trademark, brand or patent protection and are trademarks or registered trademarks of their respective holders. The use of brand names, product names, common names, trade names, product descriptions etc. even without a particular marking in this work is in no way to be construed to mean that such names may be regarded as unrestricted in respect of trademark and brand protection legislation and could thus be used by anyone.

Coverbild / Cover image: www.ingimage.com

Verlag / Publisher:
LAP LAMBERT Academic Publishing
ist ein Imprint der / is a trademark of
OmniScriptum GmbH & Co. KG
Heinrich-Böcking-Str. 6-8, 66121 Saarbrücken, Deutschland / Germany
Email: info@lap-publishing.com

Herstellung: siehe letzte Seite /
Printed at: see last page
ISBN: 978-3-659-71947-9

Copyright © 2015 OmniScriptum GmbH & Co. KG
Alle Rechte vorbehalten. / All rights reserved. Saarbrücken 2015

Table of contents

Abstract 3

1. Introduction 4
    1.1 Sprites 5
    1.2 Elves 11
    1.3 Jets 13
2. Observation and data 30
    2.1 Observation system and experiment 31
    2.2 WWLLN 33
    2.3 Local lightning location network 34
    2.4 Doppler radar 35
    2.5 MTSAT infrared images 35
    2.6 NCEP Reanalysis and radiosonde data 37
    2.7 TRMM satellite data 37
3. Results and discussions 39
    3.1 Sprite spatial extension 40
    3.2 Overall characteristics of the thunderstorms 41
    3.3 Thunderstorm radar reflectivity patterns 45
    3.4 Thunderstorm microphysical structure 51
    3.5 Characteristics of thunderstorm lightning activities 68
4. Summary 78

Acknowledgement 79

References 79

Abstract: Three summer thunderstorms in the east part of China were analyzed in detail by using data from Doppler radar, lightning location network, TRMM (Tropical Rainfall Measuring Mission), MTSAT (Multi-Function Transport Satellite) satellite images, NCEP and radiosonde. Two of them are sprites-producing storms and one is non-sprites-producing. The two sprites-producing storms occurred on 1-2 August and 27-28 July 2007, and produced sixteen and one sprites, respectively. The non-sprites-producing storm occurred on 29-30 July, 2007. The major objective of this study is try to find difference between sprites-producing and non-sprites producing storms based on multiple data set. The results show that convection in 1-2 August storm was the strongest compared with the other two storms, and it produced the largest number of sprites. The precipitation ice, cloud ice and cloud water in convection regions in 1-2 August storm were larger than that in the other two storms, but it seemed contrary in weak regions. The storm microphysical structures along lines through parent CG location showed no special characteristics related to sprites. The flash rate evolution in 1-2 August storm provides additional confirmation that major sprite activity are introduced by strong decrease in negative CG flash rate. But the evolution curve of CG flash rate is erratic in sprites-producing storm on 27-28 July, which is significantly different from that in 1-2 August storm. But it is confirmed that only lightning flashes that related to the sprite-producing cell were included in 27-28 July storm. The average positive CG peak current in sprites-producing storms were larger than that in the non-sprites-producing one. The common features of the three storms was that each storm lasted a long time (more than ten hours) and consisted of multiple convective cells, with the dissipation of previous cells, the newly formed one strengthened. The three storms have similar distributions with most precipitation ice located at 6-8 km, cloud ice at 10-14 km and cloud water at 4-5 km.

Keywords: Sprites, Doppler radar, TRMM, Lightning, Thunderstorm

1. Introduction

When talking about lightning, appearances of different lightning flashes which occur along with thunderstorms (winter or summer) will emerge in the minds of the people. Research effort on these kind of lightning can be traced back to Benjamin Franklin, hundreds of years ago. Lightning studied by Benjamin Franklin occurred in the troposphere. However, in addition to lightning, there are brief luminous phenomena which occur above the thunderstorms, and their top altitude can even reach the ionosphere.

In 1925, Nobel laureate T. R. Wilson predicted that electrical breakdown of air may be realized at mesospheric altitudes during thunderstorms (1925a, 1925b). After more than half a century on the night of 22 Sep 1989, the optical image of luminous phenomena above the storms was obtained accidently when testing a low-light television camera (Franz et al., 1990). The publication of the first image in Science triggered a flurry of ground and aircraft-based campaigns aiming to invest the characteristics of this kind of phenomenon. The phenomenon documented by Franz et al. (1990) later named as 'sprite' (Sentman et al., 1995). During the investigation of sprites, other types of luminous phenomena above the storms have been discovered, blue jets/blue starters, elves, gigantic jets. All those types of luminous events last for a very short time and are commonly referred to as transient luminous events (TLEs). Figure 1 shows the typical spatial scales of TLEs and corresponding occurrence locations above thunderstorms. Below, we briefly review the current state of knowledge about sprites and related ongoing research work.

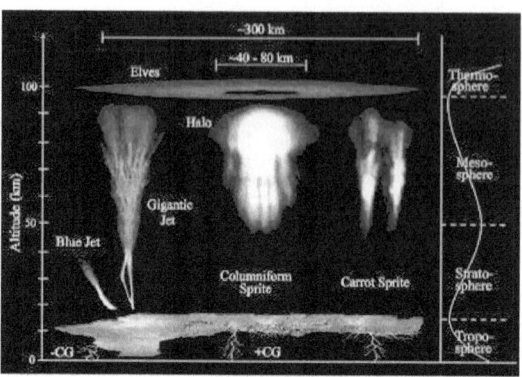

Figure 1. Several types of TLEs are shown, including sprites, halo, gigantic jet, blue jet and Elves.
Adopted from:http://www.ep.sci.hokudai.ac.jp/~msato/GLIMS/science/TLEs.html

1.1 Sprites

Sprites are the first documented and best-known upward electrical discharge. The first color image of sprites was obtained in Sprites94 aircraft campaign and the observation results established the fleeting main luminous structure to be reddish and spanning the altitude range between 50-90 km, with faint bluish tendrils which extends to downward to 40 km in altitude (Sentman et al., 1995) (Figure 2). Nowadays, sprites have been observed in north and south America, Australia, Europe, Japan, Africa, Mediterranean region and Asia (Barrington-Leigh et al., 2001; Hardman et al., 2000; Neubert et al., 2001; Hayakawa et al., 2004; Yair et al., 2004, 2009; Suzuki et al., 2006; Ganot et al., 2007; Su et al., 2002; Yang et al., 2008). Current research effort on sprites can be classified into two aspects: (1) observations; (2) mechanism. Below, we briefly review the current state of sprite knowledge and related ongoing research work.

Sprites are large luminous discharges appearing in the altitude range of 40-90 km above active thunderstorms (e.g., Sentman et al., 1995), usually coincident with a positive cloud-to-ground lightning return stroke (Boccippio et al., 1995). Sprites usually occur in cluster spanning in the horizontal range of 50-100 km, sometimes occur as single (Wescott et al., 1998). Telescopic imaging of sprites revealed that sprites have intertwined discharge channels and beads to scales down to 80 m diameter and smaller (Gerken et al., 2000; Gerken and Inan, 2002, 2003; Mende et al., 2002) (Figure 3). The dominant color in sprites is red with blue tendril filamentary structures extending downward sometimes to as low as the cloud top (Sentman et al., 1995). High speed images (1000 frames/second) show that sprite initiated at an altitude about 75 km and developed simultaneously upwards and downwards from the point of origin with an initial columniform shape, and the vertical speed is measured to be larger than 10^7 m/s (Stanley et al., 1999; Moudry et al. 2003). Higher speed (10000 frames/second) photometry reveals the sprite streamer heads (McHarg et al. 2007).

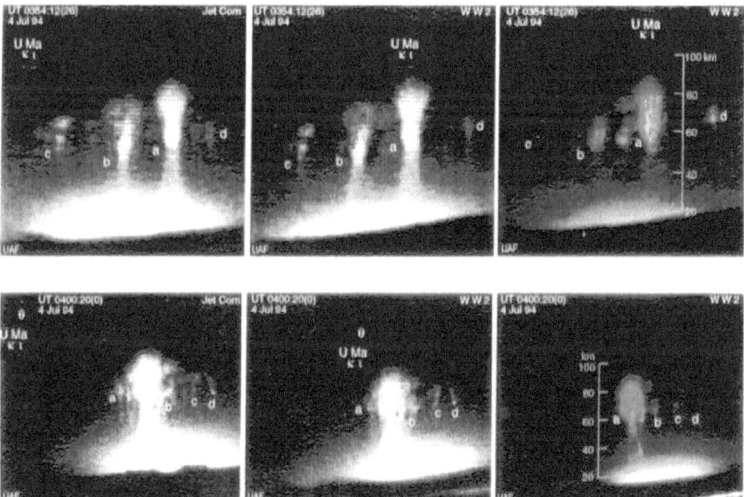

Figure 2. Color sprite images. More detailed information is available in Sentman et al. (1995)

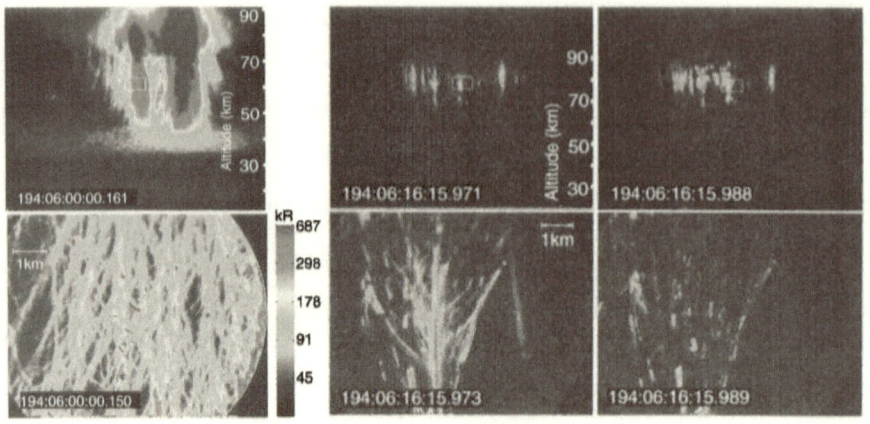

Figure 3. Wide (top panel) and narrow (lower panel) FOV images of sprite events (Gerken et al., 2000)

Spectral observations are conducted in order to investigate the physical features of sprites. Mende et al. (1995) found that the dominate emissions in sprites is the first positive band system of N_2 ($1PN_2$) (Figure 4). Almost in the same time period, the sprite spectrum was independently measured by Hampton et al. (1996). Later, the sprite blue emissions documented by Armstrong et al. (1998) and Suszcynsky et al. (1998) were found to be the second positive band system of N_2 ($2PN_2$) and the first negative band system of N_2^+ ($1NN_2^+$), which was also confirmed by Takahashi et al. (2000) and Morrill et al. (2002). The launch of the FORMOSAT - 2 satellite with scientific payload 'The Imager of Sprites and Upper Atmospheric Lightning (ISUAL)' provide a unique convenience for studying TLEs (Chern et al., 2003). The six channel photometers, covering the spectral range from the far-UV to the near infrared, provide a set of intensity data on emissions from the first positive and second positive band systems of N_2, the N_2 LBH band system, and the first negative band system of N_2^+ (Chern et al., 2003; Mende et al., 2005). Kuo et al. (2005, 2009) and Adachi et al. (2006) estimated the electric field, average and characteristic electron energy in sprites by using the photometer data (Figure 5). In addition,

TLEs occurrence rates and their global distribution have been obtained with ISUAL data (Chen et al., 2008) (Figure 6). There are also some research work on electromagnetic radiation (Cummer et al., 1998; Fullekrug and Reising, 1998; Reising et al., 1999; Stanley et al., 2000; Price et al., 2004; van der Velde et al. 2006) and infrasound emissions associated with sprites (Liszka, 2004; Farges et al., 2005; Ignaccolo et al., 2008).

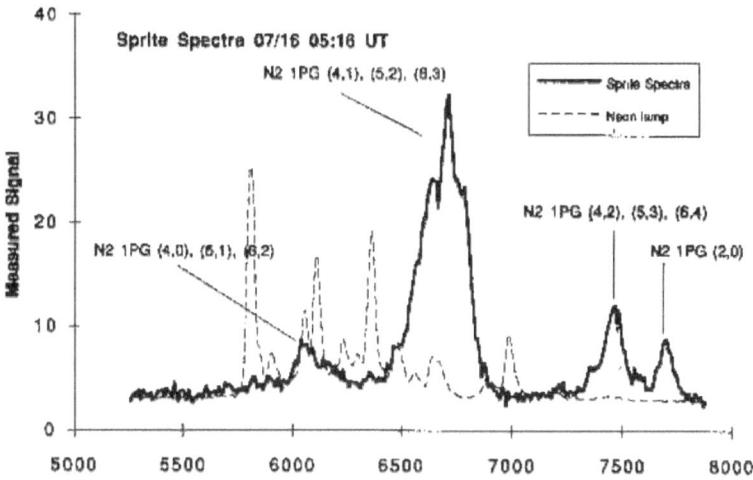

Figure 4.   Sprite spectrum obtained by Mende et al. (1995)

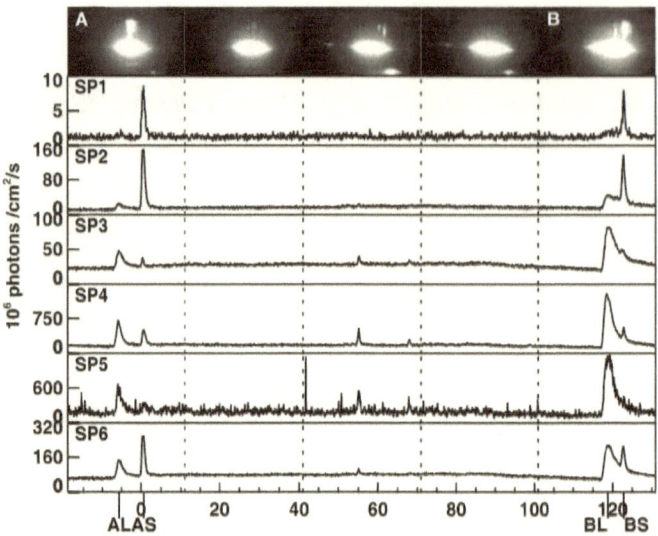

Figure 5. Spectrum of a dancing sprite (Kuo et al., 2005)

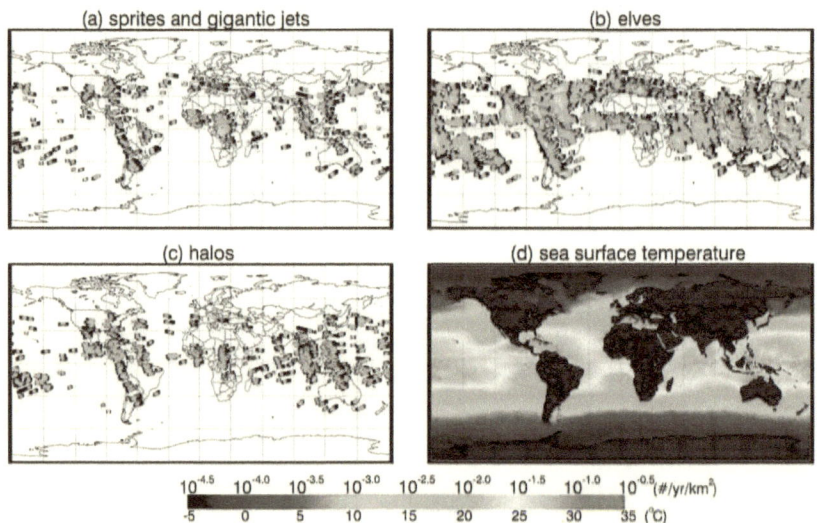

Figure 6. The global occurrence density of major TLEs: (a) sprites and gigantic jets (gigantic jets aremarked by red filled circle), (b) elves, and (c) halos. The mean sea

surface temperature between July 2004and December 2005 is displayed in Figure 3d for comparison. (Data Source: PO.DAAC, JPL) (Chen et al., 2008)

In addition to the observational work on sprites, different mechanisms for sprites production have been proposed. One is the conventional breakdown and the other is the runaway breakdown. However, along with the new results of observation, the runaway breakdown was challenged by some reasons. Telescopic imaging of sprites (Gerken et al., 2000) showed that sprites are observed to be composed of filamentary streamers which are similar to conventional breakdown. On the other hand, sprites do not follow the geomagnetic fields as would be expected for runway electron beams at high altitudes (Gurevich et al., 1996; Papadopoulos et al., 1996, 1997). Currently, the accepted mechanism for sprites is that they are an entirely conventional discharge process and the model is shown in Figure 7 (Pasko et al., 1997). Following a cloud-to-ground lightning discharges (CG), an electric field in the high-altitude regions above the thundercloud will exceed the conventional breakdown field, often around about 70 km altitude. This electric field then initiates positive streamers which grow and branch, forming sprites. This basic picture was first proposed by C.T.R Wilson in 1925 (Wilson, 1925b), and has been achieve by Pasko et al. (1997), and more detailed information could be found in the reference therein.

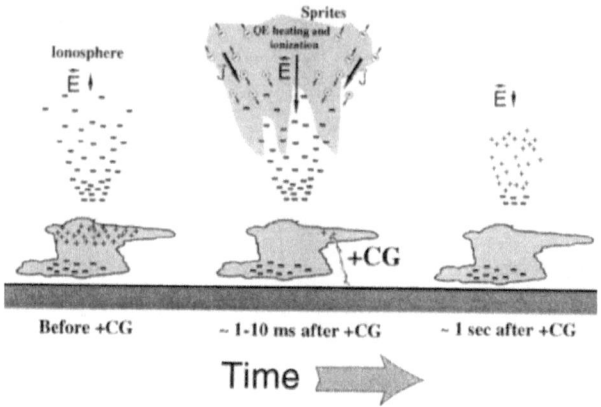

Figure 7. Illustration of the mechanism of penetration of large quasi-electrostatitc thundercloud fields to mesospheric altitudes, leading to electron heating, breakdown, ionization, and excitation of optical emissions (Pasko et al., 1997).

1.2 Elves

Elves (Emissions of Light and VLF perturbation due to EMP Sources) was first discovered by Boeck et al. (1992) in space shuttle. Later, it was confirmed by Fukunishi et al. (1996), using a multi-channel high-speed photometer and image intensified CCD cameras, it is a kind of diffuse optical flashes with a duration less than 1 ms and a horizontal scale of 100-300 km, and occur at 75-105 km altitude in the lower ionosphere just after the onset of cloud-to-ground lightning discharges (Figure 8). An array photomultiplier (Fly's Eye) designed and built in Standford University, was used to resolve the rapid lateral expansion of elves optical emission (Inan et al., 1997), and the observation results was consistent with the model predictions (Inan et al., 1991; Inan et al., 1996) in which the elves is produced as a result of heating by the electromagnetic pulse (EMP' from a lightning discharge. By using the same instrument, Barrington-Leigh and Inan (1999) confirmed that the elves had a lateral extension at least 200-700 km. Barrington-Leigh et al. (2001) by using high-speed (3000 frames per second) image-intensified video recordings and theoretical modeling, sprite halo (a brief diffuse flash sometimes observed to accompany or precede more structured sprites, which are expected to be produced by large charge moment changes occurring over relatively short timescales (~1 ms), in accordance with their altitude extent of ~70 to 85 km) and elves are successfully distinguished.

With data obtained by MEIDEX (Mediterranean Israeli Dust Experiment) instrument in space shuttle Columbia, Israelevich et al. (2004) found that elves could reach a horizontal extension exceeding 400 km, and the spatial distribution of the visible radiance indicates that they are produced by both vertical and horizontal lightning discharges. Many insights into elves have been obtained with the successful launch of FORMOSAT-2 satellite. Frey et al. (2005) showed that elves-producing lightning have some characteristics different from sprite-producing lightning. Based on electromagnetic finite difference time domain (FDTD), numerical simulation work by Kuo et al. (2007) showed that the modeled elves were in good agreement with measured results in some aspects (photon fluxes, morphology, photon counts). Kuo et al. (2012) developed a full kinetic elves model to predict its optical emissions, and the modeled emission intensity were consistent with measured results by ISUAL imager. Chen et al. (2014) comparatively analyzed the ISUAL data and World Wide Lightning Location Network (WWLLN) lightning stroke data, and found that the geographic distribution of the ISUAL elves agreed well with that for the most energetic 10% of the WWLLN lightning strokes, better than for total lighting.

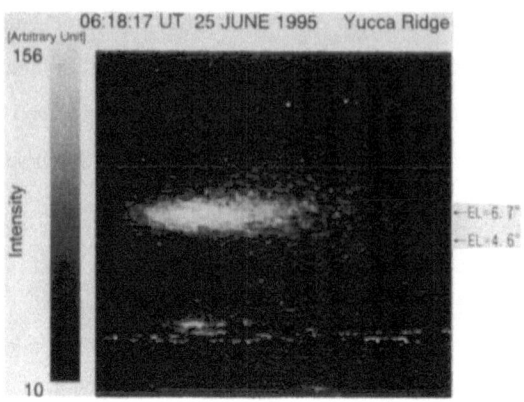

Figure 8. Elves (occurred in the lower ionosphere) recorded by Fukunishi et al. (1996). The event occurred at a distance of 537 km from the observation point, Yucca Ridge Field Station

1.3 Jets

Jets are kind of TLEs, which initiate from the thundercloud top and develop upward. According to their top altitudes, they are classified as blue starters (20–30 km), blue jets (40–50 km) and gigantic jets (70–90 km). Blue jets propagate upwards in a narrow cone from the thundercloud top at the speeds of about 100 km/s and reach terminal altitude of 40-50 km with overall luminous duration about 200–300 ms. The color of blue jet is mostly blue with some green (Wescott et al., 1995; Wescott et al., 1998; Wescott et al., 2001) (Figure 9). Blue starters are similar to blue jets, but with lower terminal about 25 km (Wescott et al., 1996; Wescott et al., 2001). The earth, atmosphere and the ionosphere consist of a sandwich structure, and the blue jets/blue starters are considered as the counterparts of cloud-to-ground lightning (CG). So relationship between blue jets/blue starters and lightning is an interesting issue. By using NLDN CG data, Wescott et al. (1996) found that within a radius of 50 km area, a sudden reduction of negative CGs persisted about three seconds after the occurrence of blue starters. They speculated that occurrence of blue jets/starters lead to the decrease of storm energy and reduction in the negative CG occurrence rate, and similar results has been found in Wescott et al. (1998). A short burst of 33 blue jets/starters occurs in a 5 min window around 12:35 UTC (22 July 2007), during the mature phase of the jet-producing thunderstorm (Chou et al., 2011).

In theory work, Pasko et al. (1996) and Pasko and George (2002) modeled blue jets/starters as streamer type discharges (resulted from the pre-discharge quasi-electrostatic field above the storm) from cloud top for a given charge distribution in the cloud. Petrov and Petrova (1999) proposed that blue jets are the streamer zones of the positive leaders, and are strongly influenced by changes in atmospheric pressure with altitude. Raizer et al. (2006, 2007) further proposed that a blue jet consists of an upward propagating leader whose top part is seen on photos as a "trunk of a tree", and is capped at the top side of the leader by its streamer zone. Later, Krehbiel et al. (2008) presented a unified view about blue jet and gigantic jets based on a combination of observational and modelling results. They found blue jets occur as a result of electrical breakdown between the upper storm charge and the screening charge attracted to the cloud top after a cloud-to-ground or intracloud discharge produces a sudden charge imbalance in the storm (Figure 10). In contrast, they found that gigantic jets begin as a normal intracloud discharge between dominant mid-level charge and a screening-depleted upper-level charges, that continues to propagate out of the top of the storm (Krehbiel et al., 2008). Riousset et al. (2010) confirmed results reported by Krehbiel et al. (2008), proposed that the accumulation of the upper screening charges near the cloudtop leads to the initiation of blue jets and modeled the mixing effect of screening charge layer and the upper thundercloud charge on the gigantic jets.

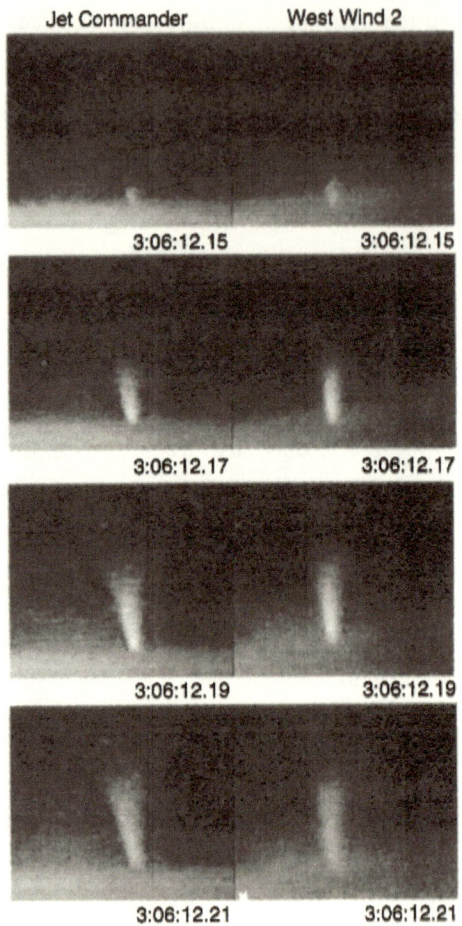

Figure 9. Time sequence of a blue jet observed from Jet Commander (left) and West Wind2 (right) (Wescott et al., 1995)

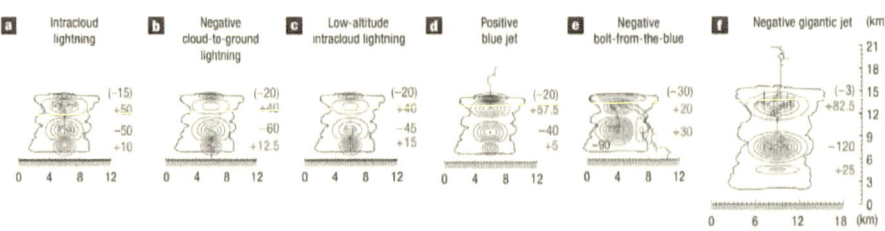

Figure 10. Simulated discharges illustrating the different known and postulated lightning types in a normally electrified storm. a–f, Blue and red contours and numbers indicate negative and positive charge regions and charge amounts (in C), respectively, each assumed to have a gaussian spatial distribution. A partially analogous set of discharges occurs or would be predicted to occur in storms having inverted electrical structures (see Supplementary Information, Fig. S5). (Krehbiel et al. 2008)

Gigantic jets are the largest one in vertical dimension in TLEs, connecting the thunderstorms and ionosphere directly and affecting the local ionosphere potential significantly. The first gigantic jet event was recorded by Pasko et al.(2002) on the evening of 15 September 2001 (taken at the Arecibo Observatory, Puerto Rico), and the discharge propagating upwards from the thundercloud to an altitude of 70~90 km, the dominated optical emission was blue (Figure 11). The electromagnetic signatures of lightning discharges (`sferics') recorded at two stations indicated that the jet itself was created by upward negative breakdown. The discovery of gigantic jet was published as cover paper in Nature. Pasko et al.(2002) did not name the observed discharge as 'gigantic jet' but as 'electrical discharge from a thundercloud top to the lower ionosphere'. Su et al. (2003) observed five similar discharges as Pasko et al.(2002) over a thunderstorm near Luzon Island, and were first named as 'gigantic jet' in their paper (Figures 12&13). Extremely-low-frequency magnetic field in four of the five events were detected, indicating that no cloud-to-ground lightning was observed to trigger these events but were produced by the gigantic jets themselves (negative cloud-to-ionosphere discharges).

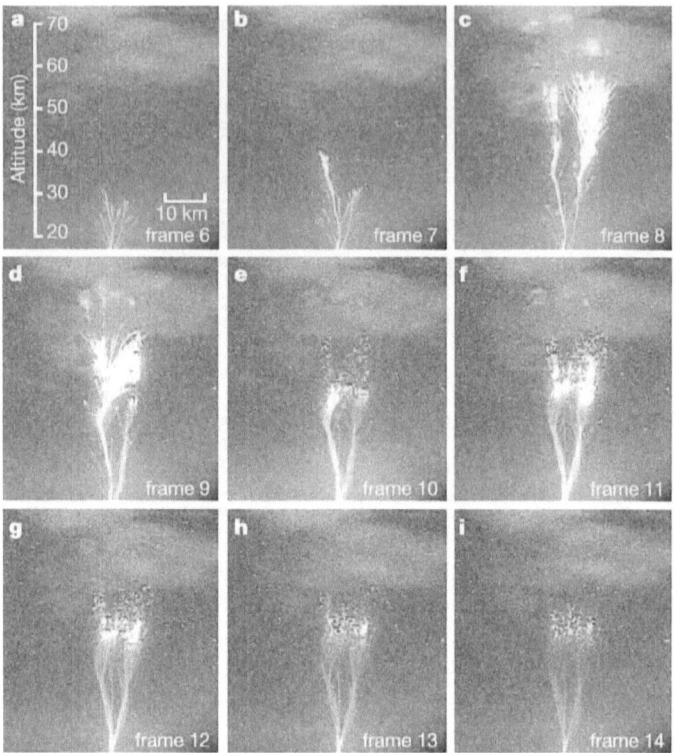

Figure 11. Sequential images of the first gigantic jet reported in Pasko et al. (2002). The video recording was obtained by using a Sony DCR TRV 730 charge-coupled device (CCD) video camera equipped with a blue extended ITT Night Vision GEN III NQ 6010 intensifier with a 40° circular field of view. The operation wavelength region of the intensifier was 390±870nm at 77% sensitivity and 350±890nm at 44% sensitivity. The intensifier provided a monochrome (predominantly green) image output. The gigantic jet event lasted a total of 24 video frames (~33 ms each), and only frames 6-14 were extracted (Pasko et al. 2002)

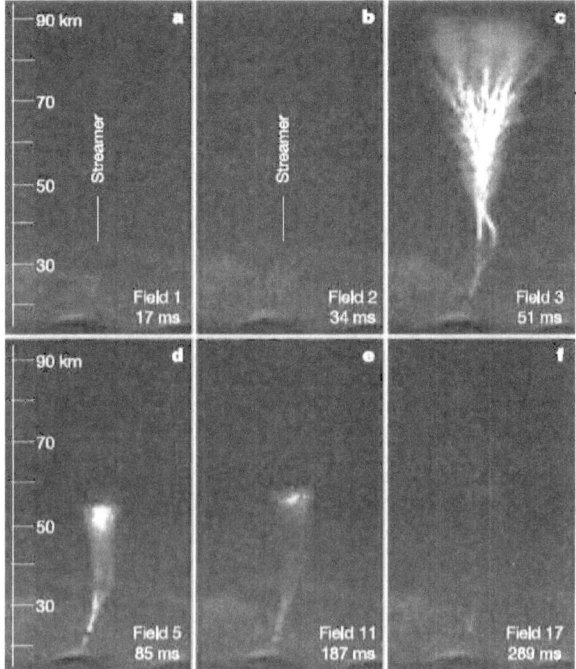

Figure 12. Gigantic jet reported in Su et al. (2003). The monochrome images were tinted to various shades of blue to bring out the salient structural features. Frame rate of the imaging system is 30 frames per second (Su et al., 2003)

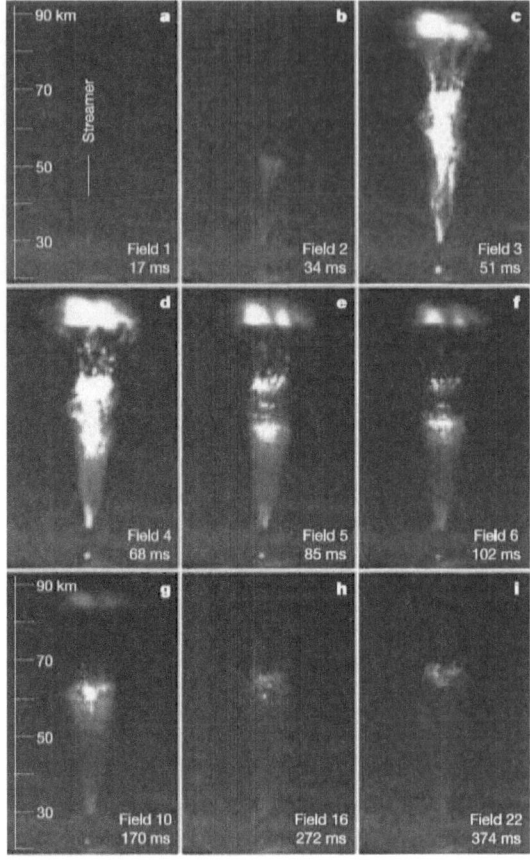

Figure 13. Another gigantic jet reported in Su et al. (2003)

Van der Velde et al. (2007) reported a gigantic jet occurred over a winter storm in continental North America and their results show that the bright lower channel of the gigantic jet ended in a fork at around 50~59 km height with very dim upper branches extended to 69~80 km. During the time window containing the gigantic jet, there was a larger and slower charge moment change of 520 Ckm over 70 ms. By using simultaneous observations of low-light video images and low-frequency magnetic field measurements, Cummer et al. (2009)reported a total charge of 144 C in a gigantic jet for the assumed channel length of 75 km, and their results confirmed the negative polarity of gigantic jets. Five gigantic jets were observed (east of Réunion Island) at a very close distance of 50 km by Soula et al. (2011) and extremely low frequency (ELF) data showed that they were all negative discharges, that is, transferring negative charges to the ionosphere (Figure 14).

Figure 14. Images of the five GJs at different stages of their development: end of the leading jet for GJ4, FDJ for GJ2 and GJ3, and beginning of the trailing jet for GJ1 and GJ5. The vertical scale is calculated at the distance of the GJ by taking into account the perspective effect and is not valid at the distance of the cloud edge (Soula et al., 2011)

Based on data obtained with ISUAL, Chou et al.(2010) classified the GJs into three types according to the generating sequence and spectral properties. Type-I GJs have similar characteristics as reported by Pasko et al. (2002) and Su et al. (2003). Type-I GJs have three stages during its evolution, the leading jet (LJ), fully developed jet (FDJ) and trailing Jet (TJ) (an example is shown in Figure 15). The leading jet acts as the stepped leader in cloud-to-ground lightning (CG), and the fully developed jet is similar to the return stroke, but the trailing Jet could not act as dart leader in CG which re-establish the complete path between the thunderstorm and ionosphere. The ELF data showed that Type-I GJs are negative cloud-to-ionosphere discharges. Type II GJs begin as blue jets (blue jets also start from thundercloud top, but their terminal altitude is about 40-50 km above the ground, lower than that of GJs) and then developed into GJs and reached the lower ionosphere (as shown in Figure 16).

Blue jets also frequently occurred at the same region before and after the type II GJs. The optical emissions of Type II GJs were weak and were inclined thought to be positive discharges. Type-III GJs were preceded by lightning, and the GJ subsequently occurred near this preceding lightning, their brightness was between Type-I and Type II GJs (see in Figure 17) (Chou et al., 2010). Chou et al. (2011) reported a Type II GJ from ground-based observation, and found that the jets occurrence may be affected by the preceding local cloud - to - ground (CG) lightning or nearby lightning (intracloud lightning or CG), while in turn the jets might also affect the ensuing lightning activity.

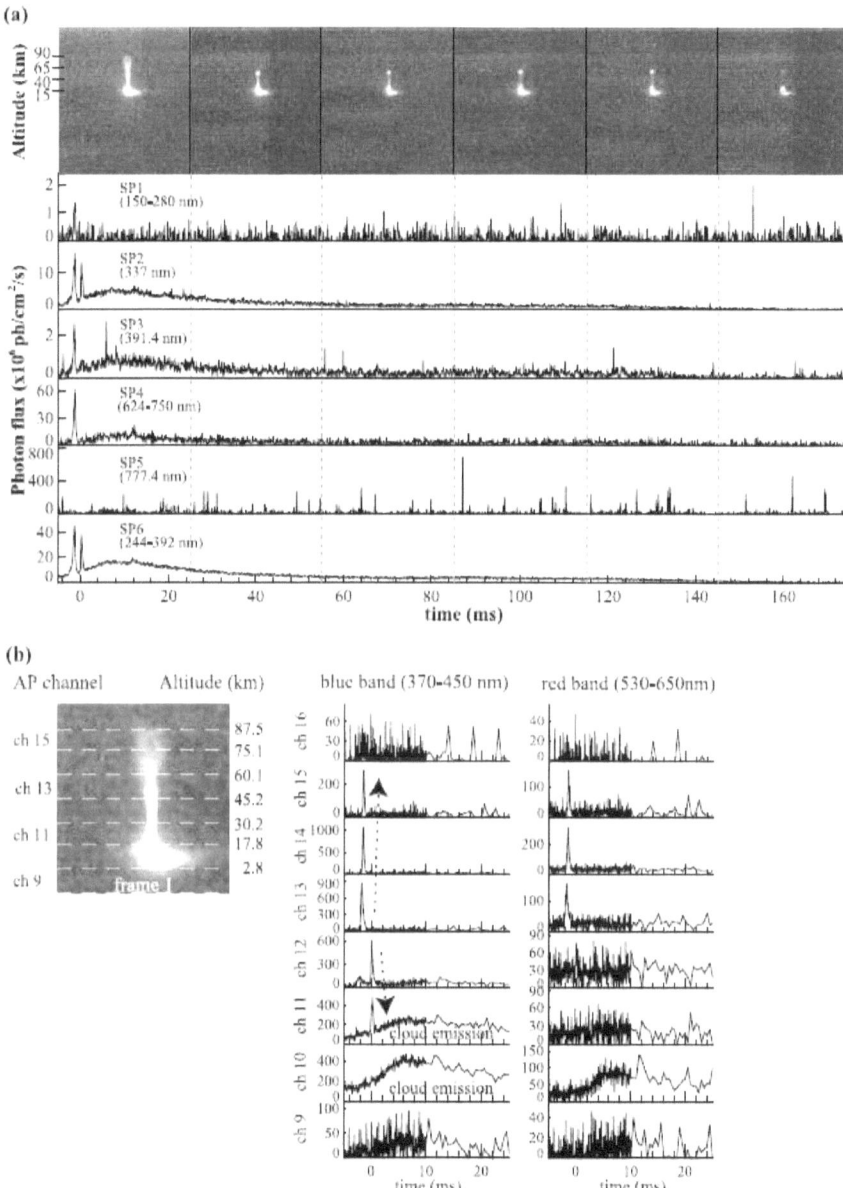

Figure 15. Type I gigantic jets (GJ) on 28 February 2006 0435:52.993 UTC. (a) The image sequence and the spectrophotometer data for this event. (b) The first image frame and the associated array photometer (AP) data. The fully developed jet

(FDJ) occurred in first image, and the corresponding spectrophotometer (SP)2 and SP6 data contain double peaks and a humping continuous luminosity. The trailing jet occurred in frames 2–6. The AP blue module shows signals associated with the upward propagation FDJ and the ensuing downward return stroke - like process. For the red module, only the signal from the FDJ was registered. The continuous cloud emissions of this event manifest themselves as the humping curves in SP2, SP3, SP6, and AP channels 10 and 11 (Chou et al., 2010)

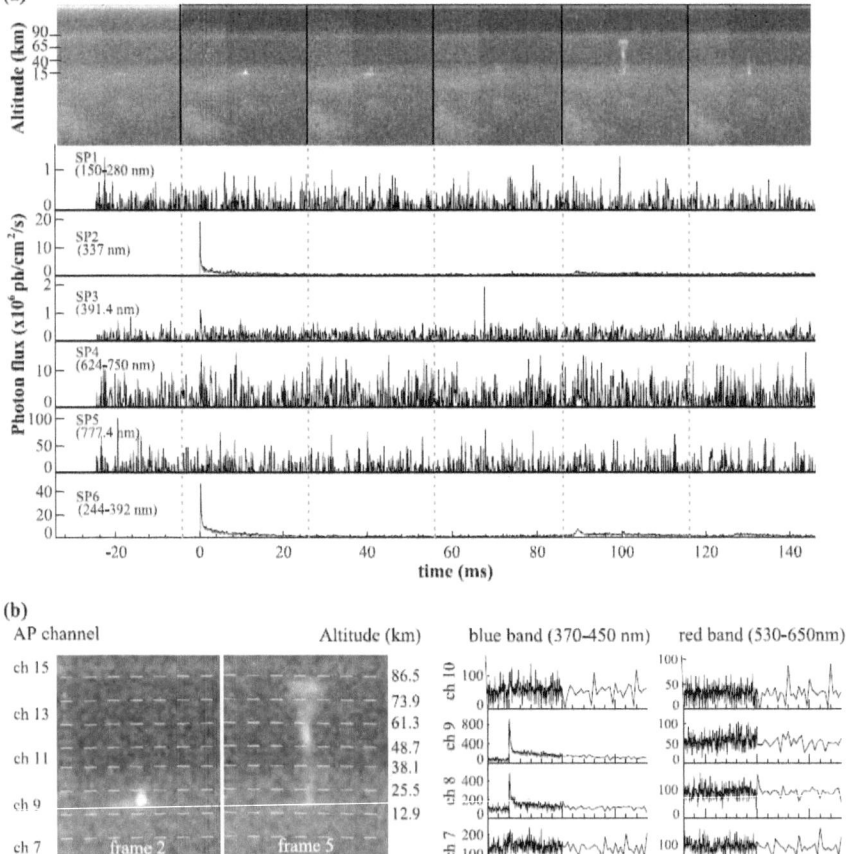

Figure 16. Type II GJ on 4 April 2009 1627:28.574 UTC. (a) The image sequence and the spectrophotometer data for this event. (b) The second and the fifth image frames and the associated array photometer data for the second frame. The blue jet occurred in the second image frame and then slowly developed into a GJ in the fifth frame. The associated AP data for the fifth frame is very noisy and is not shown here. The AP blue module data contain a clear signature from the blue jet; whereas it was absent from the AP red module data (Chou et al., 2010)

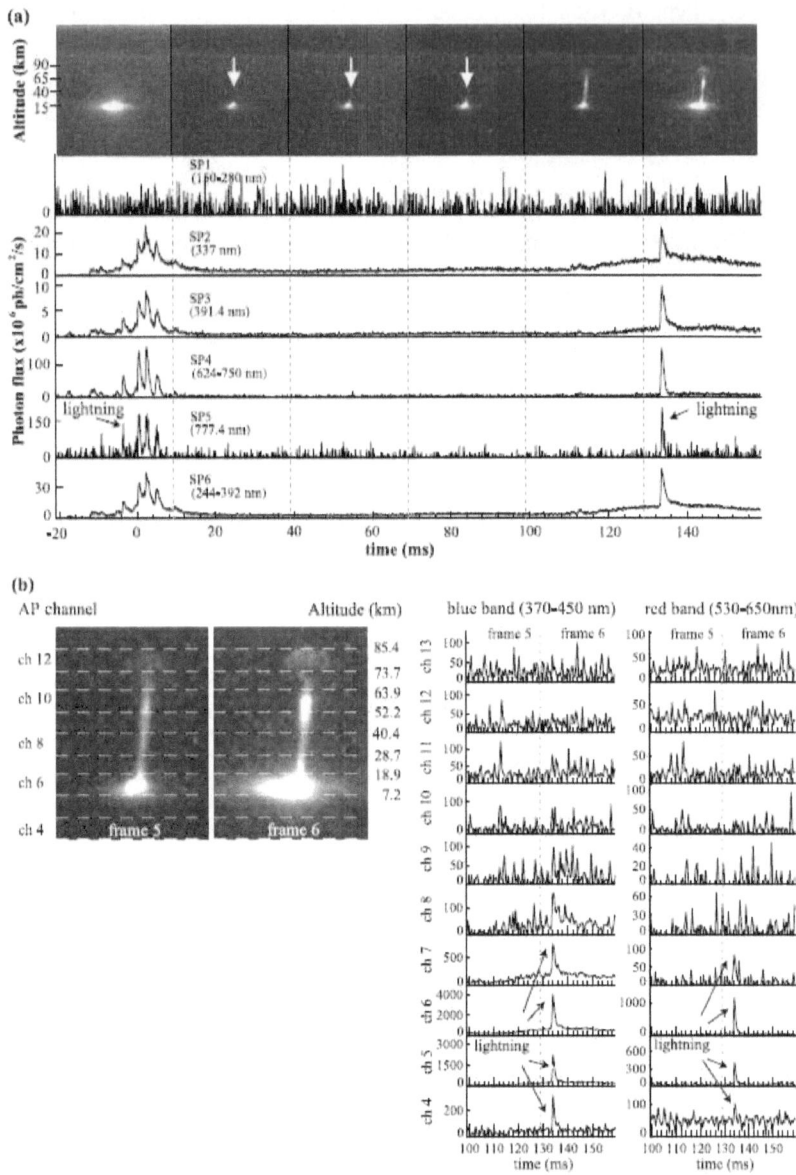

Figure 17. Type III GJ on 7 August 2005 1454:43.161 UTC. (a) The image sequence and the spectrophotometer data for this event. (b) The fifth and the sixth image frames and the associated array photometer data. The preceding lightning

occurred in the first frame. In frames 2–4, luminous columns jutting from the storm are upward discharges (marked by the arrows). The GJ occurred in the last two frames, but the SP data show no double - peaked feature during the FDJ stage. Owing to the low sampling rate (2 kHz) at this time range, the AP signals from this GJ have low signal-to-noise ratios (Chou et al., 2010)

Van der Velde et al. (2010) reported a GJ which transfer negative charge (approximately 136 C) from the ionosphere to the positively charged origins in the cloud (i.e., a positive cloud - to - ionosphere discharge, +CI), with a large total charge moment change of 11600 C km and a maximum current of 3.3 kA. The GJ occurred at 23:36:56 UTC on 12 December 2009, and reached approximately 91 km altitude, with a "trailing jet" reaching 49–59 km (Figure 18). The GJ-producing was a maritime winter type with only 6.5 km tall, showing high cloud tops are not required for initiation of GJs. Kuo et al. (2009) by using the high time resolution measurement (dual-band array photometer on FORMOSAT - 2 satellite) of gigantic jet found that the velocity of upward propagating fully developed jet stage of the gigantic jet was about 10^7 m/s, and the spectral data indicated that the electric field and average electron energy for the fully developed jet were about 400–655 Td (1 Townsend = 10^{21} V/m2) and 8.5–12.3 eV, respectively.

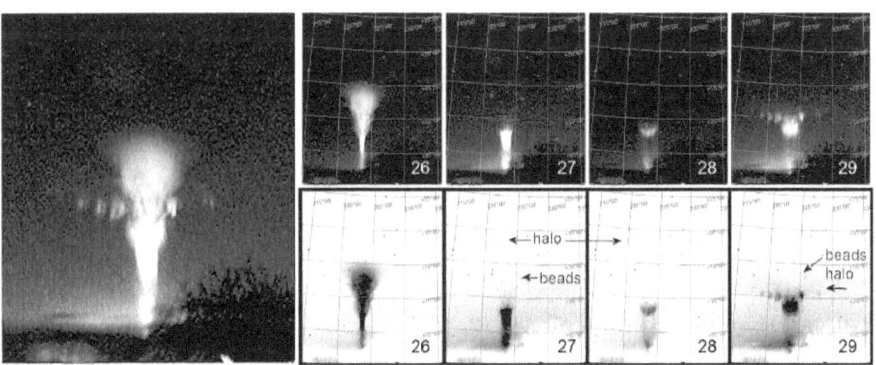

Figure 18. Sequence of video images of the gigantic jet, with frame numbers and azimuth/elevation grid. Each frame lasted 40 ms. The large image on the left is a

composite combining the brightest pixels of the video sequence. The bottom row are inverse-brightness images which show better the halo features and beads (annotated) and the diffuse light from the lightning flash. Only a part of the original wide - angle image is displayed. The azimuth - elevation grid spacing is 5°. Images courtesy of co - author Ferruccio Zanotti (Van der Velde et al. (2010))

Recently, Liu et al. (2015) reported seven jet events, including one starter, two jets and four gigantic jets in tropical region over Tropical Depression Dorian (Figure 19). The jets were captured from close-distance (about 80 km) observations, and their images were very clear. The optical and electromagnetic field data show that all events are of negative polarity. The data also indicate that the lightning-like discharge channel can extend above thunderclouds by about 30 km, but the discharge does not emit low-frequency electromagnetic radiation as normal lightning at lower altitudes (Liu et al., 2015).

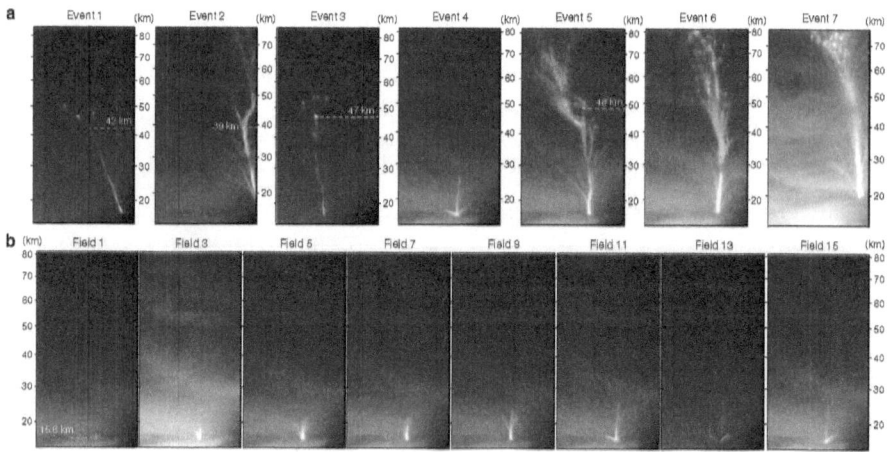

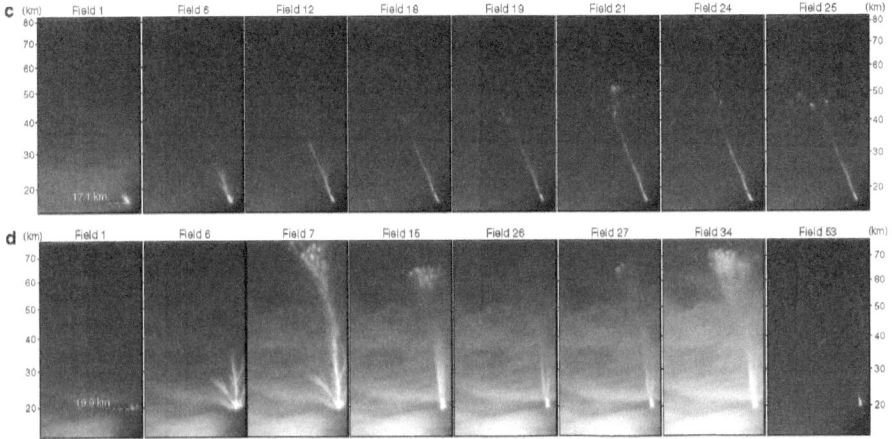

Figure 19. Low-light-level video fields of the seven upward discharges. (a) The seven events at their full extents. Events 1 and 3 are jets, event 4 is a starter and the rest of the events are gigantic jets. Selected video fields of (b) the starter (event 4), (c) a jet (event 1) and (d) a gigantic jet (event 7) (Liu et al., 2015)

Above results indicated that gigantic jets reported all occurred over tropical, subtropical, or winter storms. Compared with thousands of sprites, gigantic jet is sporadic. One GJ event was clearly recorded over eastern China (storm center located at 35.6°N,119.8°E) in Chinese mainland near the Huanghai Sea at 20:16:22 on 12 August in 2010 (Beijing time). It is the first ground-based recorded GJ that is the most distant from the equator documented over summer thunderstorm so far. The top of the GJ on image were estimated to be about 89 km (Figure 20). The GJ-producing storm was a muti-cell thunderstorm and the GJ event occurred in the storm developing stage with lowest cloud-top brightness temperature of about -73 °C and maximum radar echo top of 17 km. Altitudes with reflectivity of 45 dBZ were estimated to reach 12~14 km. Different from results in other countries that positive CGs dominated during a time period centered at GJ, negative CGs dominated during a time period centered at the GJ event and during most of the time in storm life in this storm, indicating a diversity in the lightning activity in the GJ-producing storms. It is interesting that two different storms produced two types of TLEs, that is, the GJ-producing storm only produced one GJ event during its lifetime and five sprites were produced over another storm, different from other study that sprites and GJs were produced by the same storm, enriched the knowledge of GJ-producing storms. In addition, the GJ event in their study located beyond the effective coverage area (30°S~30°N) of the ISUAL instruments onboard the FORMOSAT II satellite, and results could be considered as a useful additional reference for GJ studies.

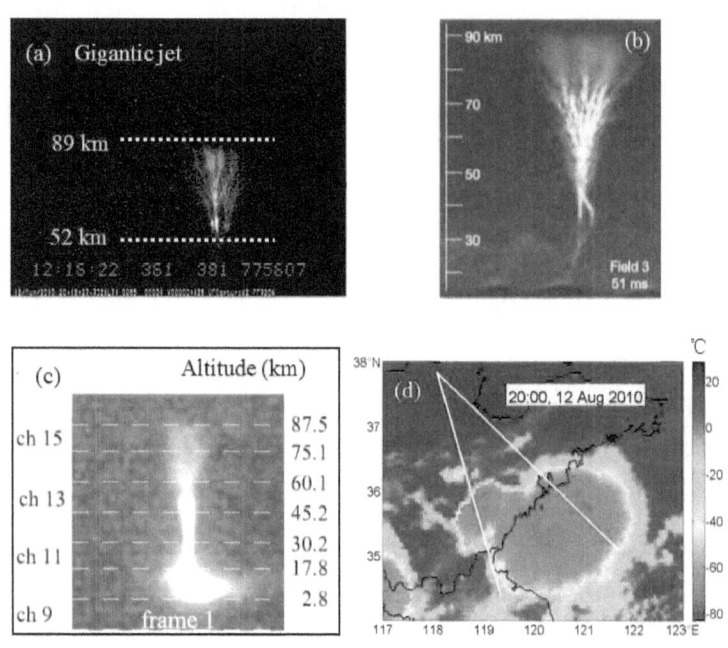

Figure 20. Gigantic jet images and its associated thunderstorm. (a) Observed GJ in Yang et al. (2012); (b) Image of Type I GJ, adopted from Su et al. (2003); (c) Image of Type I GJ, adopted from Chou et al. (2010); (d) Cloud-top brightness temperature obtained from MTSAT satellite and the lowest temperature was about -73 °C. The two white lines in Figure (d) represent the range of the line-of-sight extending from the observation site (Yang et al., 2012)

2. Observation and data

Sprites are thought to be caused by large cloud-to-ground lightning flashes. Boccippio et al. (1995) first established that a positive cloud-to-ground lightning (+CG) precedes most sprites by approximately 20-30 ms. In their study, about 86% of 42 sprites and 82% of 55 sprites were preceded by a +CG recorded by the National Lightning Detection Network (NLDN). Although Barrington-Leigh et al. (1999b) and Williams et al. (2007) found that some sprites were associated with negative cloud-to-ground lightning flashes, most studies (e.g., Lyons, 1996; Winckler, 1998; Bell et al., 1998; Soula et al., 2010) have reported results that are generally consistent with Boccippio et al. (1995).

The thunderstorms and lightning characteristicsassociated with sprites have also been studied by many authors(e.g., Sâo Sabbasa et al., 2003; Wescott et al., 2001; Cummer, 2003; Pinto et al., 2004; Marshall et al., 2007; Soula et al., 2009, 2010; Van der Velde et al., 2010; Lang et al., 2010). These studies make considerable insights into sprites, lightning and thunderstorms, and will not be elaborated in detail. Considering possible difference between sprites-producing and non-sprites producing lightning, Greenberg et al. (2007) analyzed the ELF transients of the lightning discharges which generate TLEs in the time and frequency domains, and compared them with other lightning discharges occurring in the same thunderstorm cell at approximately the same time but which did not produce TLEs. However, Greenberg et al. (2007) failed to find any significant difference. Although charge moment change is an important factor in determining if TLEs occur, it is not the only factor, other factors (for example, storm meteorological characteristcs, lightning activity, storm electrical structure) may also play roles in intiating the mesophoric electric breakdown when sprites occur.

Aiming at the investigation of possible difference between sprites-producing and non-sprites-producing thunderstorms, comparative analysis was made in detail by using multiple data including lightning location data, MTSAT (Multi-Function

Transport Satellite) satellite images, Doppler radar, TRMM (Tropical Rainfall Measuring Mission), NCEP and radiosonde. Especially detailed microphysical structure of the storms were given by TRMM. Although this is a case study, to the best of our knowledge the microphysical parameters from the TRMM satellite have not been analyzed before for sprites-producing and non-sprites producing storms except the work reported by Sâo Sabbasa et al. (2010) in which TRMM data were analyzed in a TLEs-producing storm in Brazil. Analysis in this study focus on TRMM and Doppler radar figures that show almost the same time and geographic area. Three storms (two sprite-producing storms, and one non-sprite-producing storm) have been comparatively analyzed in detail in this study. The two sprite-producing storms occurred on 1-2 August and 27-28 July 2007, respectively, and the non-sprite-producing storm was 29-30 July, 2007. This kind of analysis may be helpful for understanding storm characteristics. In addition, a variety of information of sprites-producing and non-sprites-producing storms were given by multiple data. Below we will briefly introduce the experiment and the data used in this study.

2.1 Observation system and experiment

Sprite experiment was first conducted in the eastern coast area of China (37°49'42"N, 118°05'06"E, Shandong province) during the summer of 2007, and sprite images were first captured in this year. The observation system used in the experiment is shown in Figure 21. The camera used in this experiment consists of Watec902H CCDs (PAL camera, with 1/2 inch Interline transfer CCD image sensor) and 12 mm, f/0.8 lens, and the field of view (FOV) of the observation system is 31.1° (H) by 21.2° (V). The camera is developed based on super high-sensitivity monochrome CCD, with minimum illumination of 0.0003 lux at F1.4. Data from the camera were recorded in real time in the computer hard disk, and the time of the system was synchronized to the universal time (UT) with GPS. (Yang et al., 2008).

In order to obtain more data as possible, the gain of the camera is set to high, that is, minimum illumination of 0.0003 Lux. The frame rate of the observation system is set to 25fps (25 frames per second), with a frame duration about 40 ms. The camera pointing of the observation system is determined by using the background star field (Suzuki et al., 2011). The camera is enclosed in a housing to protect it from weather's hazards and is mounted on an antenna rotator. The camera is controlled remotely from the computer via the Internet, allowing azimuth adjustment during the observation in real time. In the following years, several other stations have been setup and their outdoor cameras are shown in Figure 22. In these stations, the camera used is Watec902H2 Ultimate, and some of the stations used a 3.6 mm F1.4, leading to a field view of 100° (H) by 77° (V). Many beautiful sprite images have been obtained during these years of observation (see Figure 23).

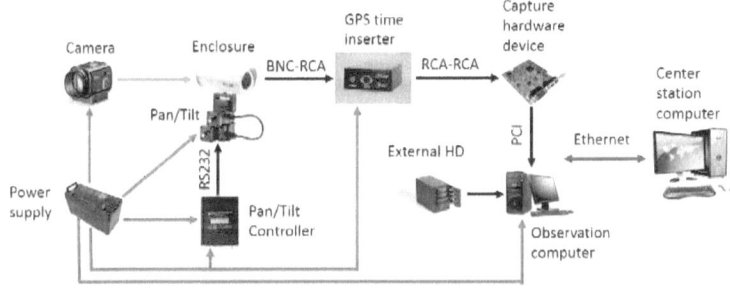

Figure 21. Diagram of the sprite observation system

Figure 22. TLEs stations setup in China

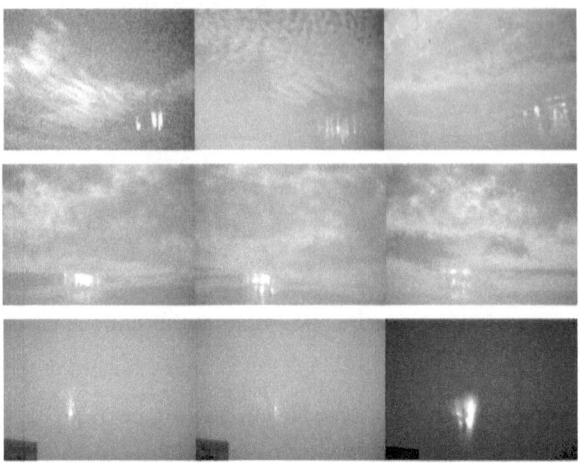

Figure 23. Some sprite images obtained during the observation experiment

2.2 WWLLN

During the experiment, infrared weather maps (MTSAT satellite, FY-2 satellite) were used to obtain the location of potential thunder-storms for observations. Based on real-time lightning information from ground lightning location network, accurate position of the center of the electrical activity was determined, which was used to adjust the camera pointing direction. The World Wide Lightning Location Network (WWLLN) is used for this purpose.

WWLLN is a VLF lightning antennas array, more than fifty antennas located around the world, of which one antenna is located at the Institute of Atmospheric Physics, Chinese Academy of Sciences (distributed by the University of Washington in Seattle). Each lightning stroke location requires the time of group arrival (TOGA) from a least 5 WWLLN sensors. After processing the data with some method, the system provides global lightning strokes, and global map with lightning strokes positions is updated every 10 minutes, and another map with a superimposed cloud image viewed by satellite (of the National Weather Service/Aviation Weather Center) are updates every 60 minutes. Figure 24 is an example of the map, showing a WWLLN map with superimposed satellite image. The blue dots are the lightning detections accumulated in the hour before image, circled Red asterisk are the locations of the WWLLN antennas, and the Red line represents the terminator. For reanalysis the system provides time and location of each lightning strokes (latitude and longitude) and the number of stations which detected the strokes. The system does not provide electrical parameters of the strokes such as peak current and charge-moment changes, etc.

Figure 24. WWLLN map with superimposed satellite cloud image

2.3 Local lightning location network

In addition to identify lightning activity in real time, another lightning location data are used for reanalysis of lightning characteristics that produced the TLEs. The lightning location network which consists of 10 sensors and one data processing center, and uses a combined TOA/MDF technology (Cummins et al., 1998). Each cloud-to-ground lightning record comprises data on the occurrence time, location, peak current, and polarity of lightning. The location accuracy of the network is about 2000 m, which has been proved by the fault of high-voltage transmission line (Liu et al., 1997). The detection efficiency of the lightning location network is about 90% (Feng et al., 2007). In addition to lightnig data given by Shandong meteorological bureau, similar lightning data is given by Henan meteorological bureau.

2.4 Doppler radar

The thunderstorm evolution and structure characteristics are given by Doppler radar data which is a useful tool for monitoring the mesoscale convective systems (MCSs). Radar used in this study is aWSR-98D S-band Fully Coherent Doppler Weather radar, and it is an upgraded version of WSR-88D. It has many features including state-of-the-art computerized control, 24-hour operational capability, real-time monitoring, real-time calibration, and high accuracy and reliability. The radar can operate in two modes, clear-air and precipitation mode. It has two scanning range, one is 230 km and the other is 460 km. In order to get accurate information, scanning of 230 km (with a resolution about 1 km) is used in this study. The radar image is updated every 6 minutes. The data was provided by Shandong Meteorological Bureau and has been used to characterize the structure and evolution of the storm.

2.5 MTSAT infrared images

MTSAT satellite data are used to supply cloud rough characteristics every hour with spatial resolution of 0.05°×0.05°, and was downloaded from website (http://weather.is.kochi-u.ac.jp/archive-e.html) for reanalysis. The MTSAT was launched in 2005 and is three-axis stabilized. Its geostationary orbit is 35,800 km above the equator at 140° east longitude or 145° east longitude. The MTSAT carries five sensors, that is VIS (0.55–0.90 μm), IR1 (10.3–11.3 μm), IR2 (11.5–12.5 μm), IR3 (6.5–7.0 μm) and IR4 (3.5–4.0 μm), information could be found in http://www.jma.go.jp/jma/jma-eng/satellite/about_mt/8.Major_Characteristics_of_the _meteorological_payload.html (Figure 25). Data used in this paper is IR1. In addition to MTSAT satellite data, infrared images from FY-2 satellite is also used for adjusting the camera pointing of direction (Figure 26).

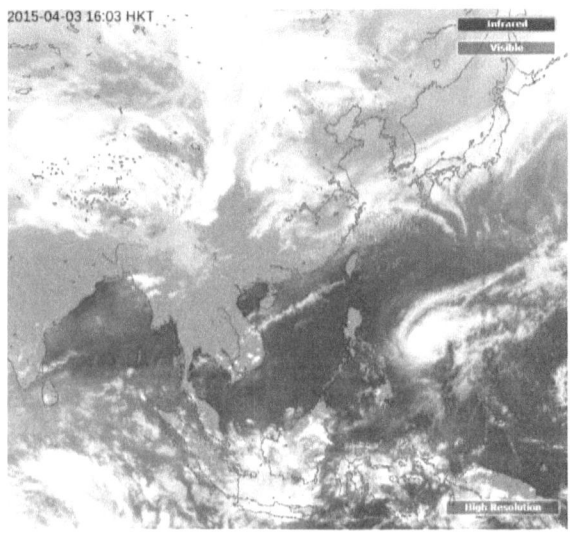

Figure 25. An example of MTSAT infrared image
(http://www.hko.gov.hk/wxinfo/intersat/satellite/sate.htm)

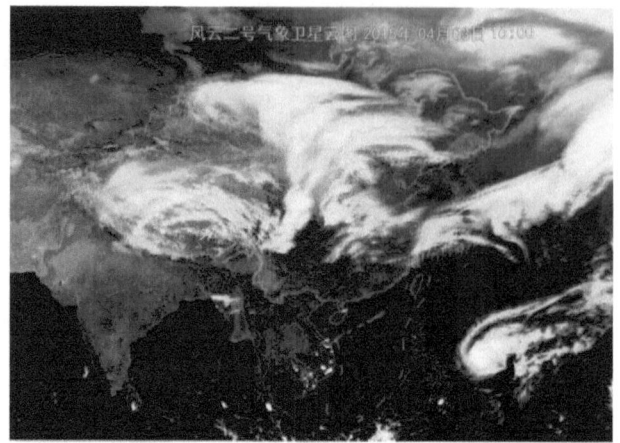

Figure 26. An example of FY-2 infrared image
(http://www.hko.gov.hk/wxinfo/intersat/satellite/sate.htm)

2.6 NCEP Reanalysis and radiosonde data

The NCEP Reanalysis Derived data was downloaded from website (http://www.cdc.noaa.gov/) and processed via GrADS software. The data is given in 4-times daily(00Z, 06Z, 12Z and 18Z). Radiosonde data from Qingdao station (215 km away from the 1-2 August storm, 444 km away from the 27-28 July storm) were used for storms on 1-2 August and 27-28 July, and were downloaded from Wyoming University website (weather.uwyo.edu/upperair/sounding.html). Zhengzhou radiosonde data have been used for storm on 29-30 July, and it located 222 km west of the storm. So naturally accuracy of the radiosonde data is limited. Nevertheless, these radiosonde stations were the closest we could find.

2.7 TRMM satellite data

The TRMM (Tropical Rainfall Measuring Mission) satellite launched jointly by the United States and Japan in 1997, carries five detectors, the Precipitation Radar (PR), the TRMM Microwave Imager (TMI), the Lightning Imaging Sensor (LIS), the Visible and Infrared Scanner (VIRS) and the Clouds and the Earth's Radiant Energy System (CERES) (e.g. Kummerow et al., 1998). The PR is an electronically scanning radar, operating at 13.8 GHz that measures the three dimension rainfall distribution over both land and ocean, and define the layer depth of the precipitation. TMI is a multi-channel dual-polarization passive microwave radiometer and utilizes nine channels with operating frequencies of l0.65 GHz, 19.35 GHz, 21.3 GHz, 37 GHz, and 85.5 GHz (21.3 GHz just for vertical polarization), providing information on the integrated column precipitation content, cloud liquid water, cloud ice, rain intensity, and rainfall types (e.g., stratiform or convective). LIS is a calibrated optical sensor operating at 777.4 nm, and it observes the distribution and variability of lightning over Earth. The VIRS is a five-channel, cross-track scanning radiometer operating at 0.63, 1.6, 3.75, 10.8, and 12 um, which provides high resolution observations on cloud coverage, cloud type, and cloud top temperatures. CERES is a broadband scanning radiometer with a total spectral range of 0.3 to 50 um; it measures emitted and reflected radiative energy from Earth's surface and the atmosphere and its constituents (e.g., clouds, aerosols).

Although this satellite passes above the research area only twice a day, the three storms were fortunately scanned. Because the swath width of Precipitation Radar (PR) (247 km) is much smaller than that of TRMM Microwave Imager (TMI) (878 km), only TMI could be used for analysis (the PR radar only covered a small part of some storms, and is not appropriate for analysis). Data 2A12 of TMI has been used in this study. In addition, 1B11 data is also used for analyzing the polarization corrected temperature of the storm. More information on TRMM data could be found in http://trmm.gsfc.nasa.gov/.

3. Results and discussions

The sprite experiment lasted two months, and a total of seventeen sprites were observed above two thunderstorms. Sixteen sprites were observed at the late night of August 1 and in the early morning of August 2 (local time is used in this paper). One sprite was observed over thunderstorm on 27-28 July. There was fog in the camera FOV in the late stage of the 27-28 July storm (this storm moved towards the observation site during its evolution), it is not sure whether this storm produced sprites or not after the only recorded sprites at 00:36:39 28th July. Anyway, because one sprite was recorded during 27-28 July storm, it is considered as a case of sprites-producing storm. Therefore, there are two cases of sprites-producing storms, one is the 1-2 August storm and the other is the 27-28 July storm. Although Chen et al. (2008) found that 80% of the TLEs recorded from July 2004 to June 2007 were elves, no elves were observed during our experiment. It is possible that no elves occurred, and it is also possible that elves occurred and could not be seen by the camera because of the cloud in the camera field view. Considering the non-sprites-producing storm, the storm at the late night of 29th and in the early morning of 30th July in 2007 has been analyzed.

All of the three thunderstorms lasted a long time (more than ten hours) and were mesoscale convective systems (MCSs), similar to other studies (Lyons et al., 2003; Soula et al., 2009). With the older cells dissipating the new cells were developing and became strong. Two orbit data could be used for thunderstorms on 1-2 August (orbit number 55325 and 55326) and on 29-30 July (orbit number 55279 and 55280). Since the paper focus on sprites, and most sprites occurred in the storm mature-to-dissipating stage, orbit data that scanned this stage were fully used. In addition, analysis focus on TRMM and Doppler radar figures that show almost the same time (which could be seen in next sections) and geographic area. This kind of analysis may be helpful for understanding storm characteristics.

3.1 Sprite spatial extension

Sprites occur over thunderstorms, but what are their spatial dimensions? To derive this information one has to know the distance between the sprites and the observation site. Since there was only one observation site in the campaign, the location of the observed sprites could not be accurately determined. Nevertheless, sprites usually appear in the vicinity (about 50 km) of their parental cloud-to-ground lightning flashes (CGs) (Sabbas et al., 2003). The distance of the CGs from the observation site could be deduced. With the above information and the FOV of the observation system, the heights and extensions of the sprites could be estimated using the method reported by Hsu et al. (2003). Using this method, the estimated sprite tops shown in Figure 27&28 were about 85 km, similar to the results reported by Hsu et al.(2003).

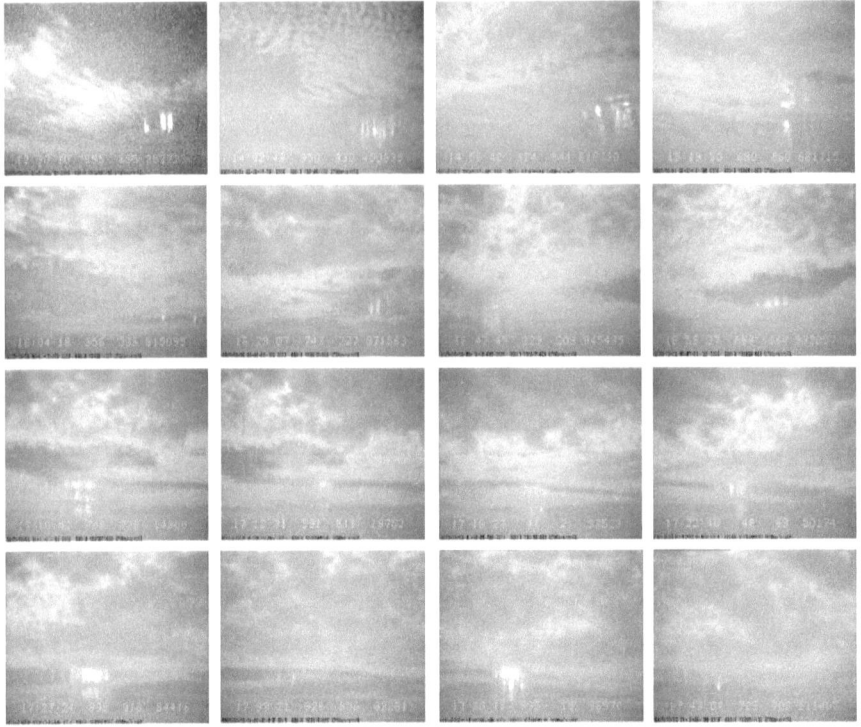

Figure 27.   TLEs captured over storm on 1-2 August 2007.

Figure 28.   Sprite captured over storm on 27-28 July 2007.

3.2 Overall characteristics of the thunderstorms

The meteorological data, including cloud top temperatures and height, sea surface temperature and CAPE values are summarized in Table 1. Sea surface temperature data for the Yellow Sea were taken from NCEP/NCAR reanalysis, and were found to be about 26℃ during the three observation days. The cloud top heights and areas of parent active region were obtained from Doppler radar. The cloud top brightness temperature was obtained from the MTSAT satellite images. The radiosonde data were used to get the convective available potential energy (CAPE), which is a valuable parameter to indicate the thunderstorm development. The locations of the three storms and radiosonde stations were shown in Figure 29.

Table 1.   Meteorological characteristics in the three storms

Date	1-2, August	27-28, July	29-30, July
Parent active region (km^2) (with reflectivity larger than 30 dBZ)	14800	6903	17993
Cloud Top Temp (°C)	-50	-43	-55

Cloud Top Height (km) (with reflectivity larger than 30 dBZ)	9-11	6-8	6-8
Sea Surface Temp (°C)	26	26	26
CAPE (J/Kg)	1703-551	2388-1960	3197-485
Number of TLEs observed	16	1	0

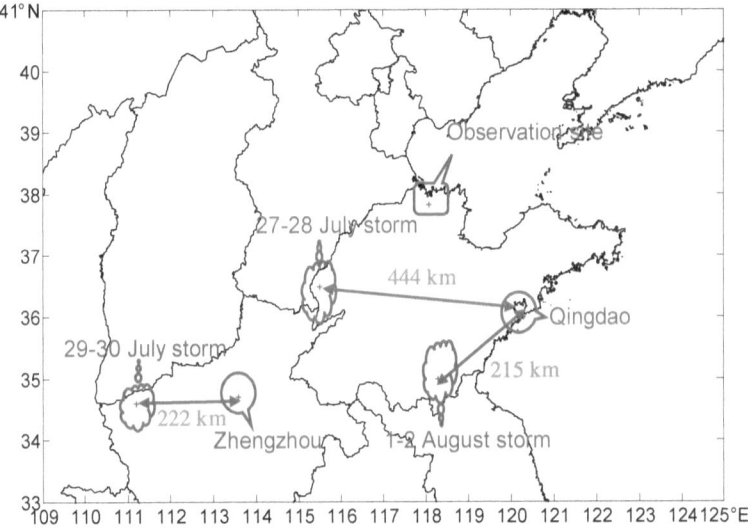

Figure 29. Locations of the observation site, thunderstorms, and radiosonde stations.

The CAPE values in the two sprite-producing days were larger than 1000 J/kg (20:00 27th July, 20:00 1st August) during the storm evolutions. The CAPE values at 02:00 28th July and at 02:00 2nd August were about 1960 J/kg and 551 J/kg, respectively, indicating that the CAPE value was more consumed in 1-2 August storm than that in 27-28 July storm. Considering the non-sprites producing storm on 29-30 July, its CAPE value was larger than that of the two sprites-producing storms. In addition, CAPE value at 02:00 30th July was smaller than that of the two sprites-producing storms. CAPE value in 29-30 July storm was consumed by a large amount of lightning flahses, which could be seen in the next sections (This storm produced the largest number of flashes). Since most sprites were produced during the storm mature-to-dissipating stage, active regions with radar reflectivity larger than 30 dBZ in this stage was estimated in the three storms (see Table 1). It is found that parent active regions in 1-2 August storm was much larger than that in 27-28 July storm. But regions with reflectivity larger than 30 dBZ in 29-30 July storm was the largest in the three storms. Cloud top heights with 9-11 km dominated in 1-2 August storm, but cloud top heights with 6-8 km domimated in 27-28 July and 29-30 July storms.

All of the three storms lasted more than ten hours with the minmum of 12 hours in 27-28 July storm and maximum of 16 hours in 1-2 August storm. The 29-30 July storm lasted 13 hours. Figure 30 shows cloud-top brightness temperature, CG flash, and parent CGs distributions in the three storms. The hours shown in Figure 25 corresponded with the storm mature-to-dissipating stage and it was at night in the observation site. Figure 30 shows the cloud top brightness temperature evolution associated with lightning and sprites during the time period chosen. The results show that as the storms grows, negative CGs more than positive CGs concentrated in compact regions with very cold cloudtops. Not all regions with very cold cloudtops exhibited lightning activity. But charge separations are expected follows convection (very cold cloudtops usually corresponded with strong convections), leading to intensification of lightning activity, which could be seen at 02:00 on 2nd August storm. The lightning flash rate increased after 02:00 on 2nd August (see the lightning flash rate evolution in Figure 43). Similar phenomenon could also be seen in 27-28 July and 29-30 July storms (02:00 28th July, 04:00 30th July). The parent CGs that was found located in regions with cloud top brightness temperature of -40 ℃~ -60 ℃, similar to result (-45 ℃~ -53 ℃) reported by Sâo Sabbasa et al. (2010) and result (-50 ℃~ -55 ℃) reported by Soula et al. (2009). Positive CGs were not active in the region with high negative CG flash density in some hours, but this is not always the case in all the hours in Figure 25. Both of positive and negative CGs occurred in cold cloudtop temperature regions in some hours.

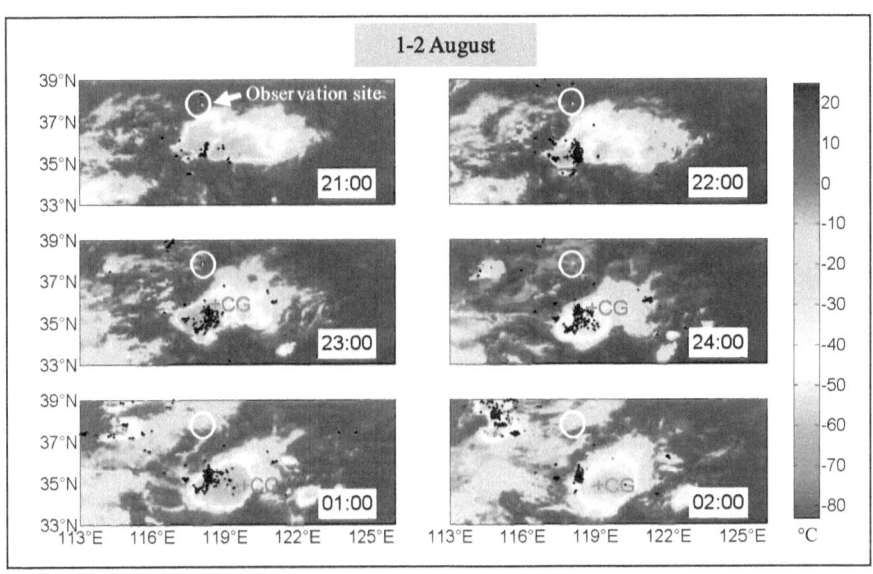

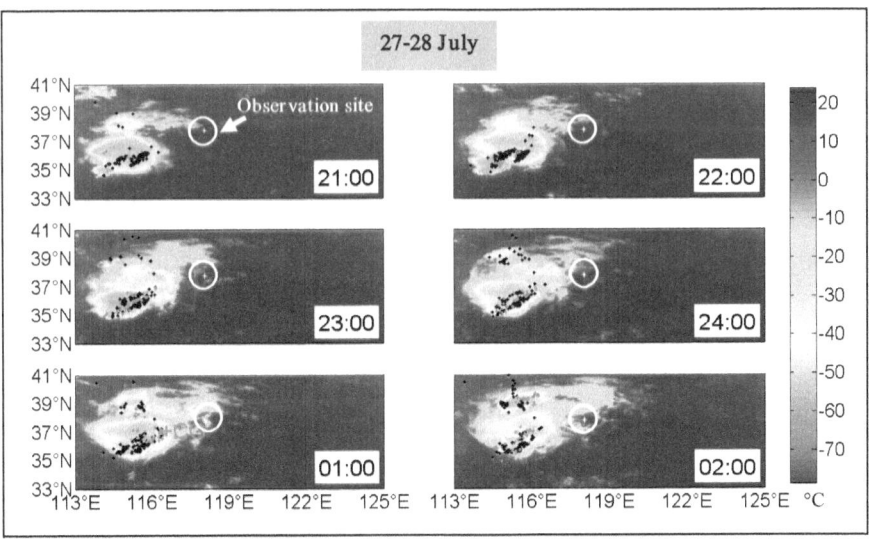

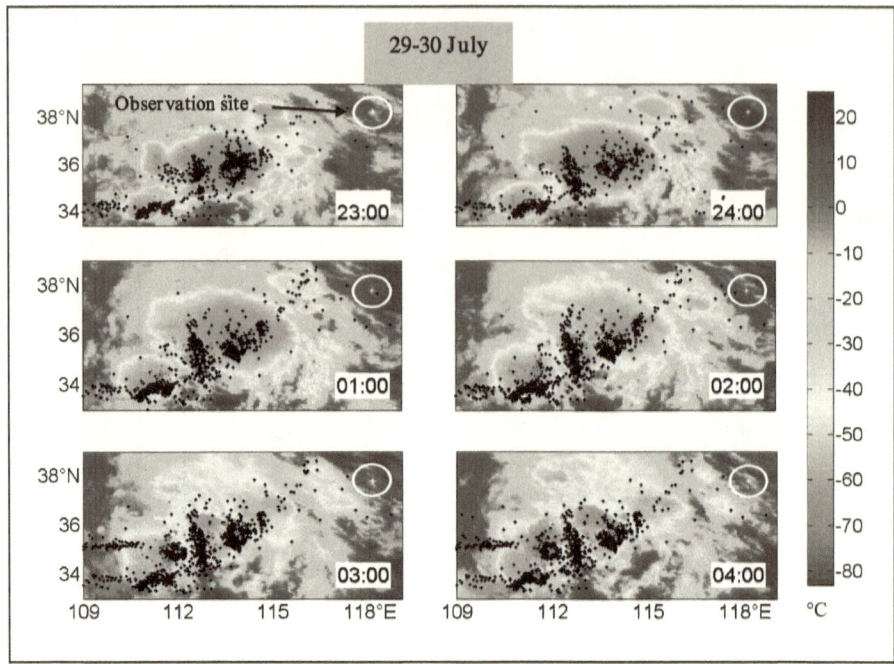

Figure 30. Cloud-top brightness temperature with CG flashes within an hour (±30 min) centered at the time shown on the figure. Red rose dot and black dot in the figure stand for positive and negative CG lightning, respectively. Symbols '+' stand for parent CGs. The circled white asterisk shown on the image indicates the observation site

3.3 Thunderstorm radar reflectivity patterns

The resolution of MTSAT infrared weather images is rough, the storm detailed information will be given in this section by Doppler radar. Since this paper focus on sprites, and most sprites occurred in storm mature-to-dissipating stage, Doppler radar data in this stage will be fully used. Sixteen sprites were recorded over 1-2 August thunderstorm, different from the storm reported by Lang et al. (2010) which produced 282 TLEs during 4 hours, also different from study by Sâo Sabbasa et al. (2010) in which one storm in Argentina produced 444 TLEs. Figure 31 shows that the 1-2 August storm developed at 15:00 on 1st August (storm C1), and became strongest at 17:30 with maximum composite radar reflectivity of 50 dBZ. Then this storm weakened, and new small storm (storm C2) occurred at 21:00 and developed very rapidly with maximum composite radar reflectivity of 55 dBZ at 22:00 (coldest cloud-top brightness temperature of -66.96 °C). Since the parent CG of the first sprite on 21:30:10 was not found, it was not known which storm (C1 or C2) produced this sprite. The cloud coverage of storm C2 at 23:00 (storm C1 almost disappeared at this time as shown in Figure 31) became larger compared with that at 22:00, and the fourth sprite was recorded at 23:19:50, and its parent CGs located in stratiform region with radar reflectivity of 25 dBZ (the large stratiform regions corresponded with comparatively warm regions compared with the coldest temperature of -64.12 °C at 23:00). All of the parent CGs that were found located in stratiform regions with reflectivity of 15-30 dBZ ('+CG' in the figure was parent CG, the storm motion is also shown in the figure), in good agreement with study reported by Soula et al. (2009, 2010). Liu et al. (2011) also found that positive CGs occurred in stratiform region in a leadling-line and trailing stratiform mesoscale convective system in this region. It should be noted that parent CGs shown in figures labeled '23:17' and '23:35' were the same. Figure 31 shows that radar reflectivity in figures labeled '23:17' and '23:35' had similar characteristics. One can expect that vertical structures along GH and MN in figures labeled '23:17' and '23:35' also had similar characteristics. Therefore, although TRMM passed the storm at 23:34, the vertical structure along GH and MN at this time could show the storm characteristics at 23:17. The storm vertical structure along the parent CG (along lines GH and MN shown in the figure labeled '23:35') will be given by TRMM in the next section.

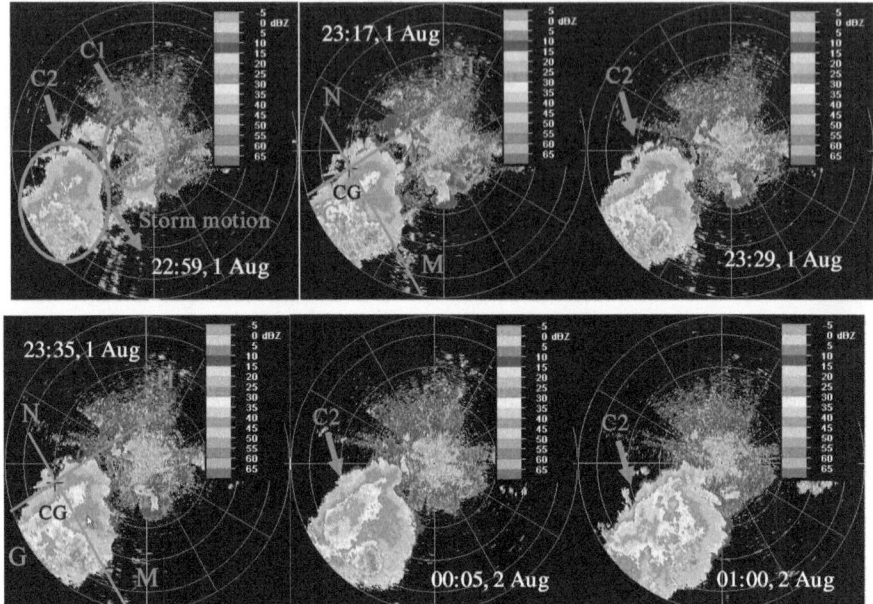

Figure 31. Radar reflectivity of the storm on 1-2 August obtained from the Doppler radar data. The time and storm motion are also shown in the figure ('+' in the figure labeled parental CG). Parent CG in '23:35' and '23:17' images is the same

One sprite was recorded over thunderstorm on 27-28 July. The MTSAT infrared weather maps showed that this storm occurrred at 16:00 27th July, but not in the scanning range of the Doppler radar at this time. At 19:00, Doppler radar could 'see' this storm (storm C1 in Figure 32), and its maximum radar reflectivity was 50 dBZ. At 21:05, storm C1 weakened and new storm (named C2) occurred in the southwest of storm C1 and developed rapidly. Storm C1 became more weaker and C2 became strong, and the two storms almost connected together at 23:02. One columniform sprite was recorded at 00:36:39 and its parent CG located in stratiform regions with radar reflectivity of 30 dBZ ('+CG' in Figure 32), and C1 almost disappeared at this time. Similar to storm on 1-2 August, large stratiform regions also occurred during 27-28 July storm evolution. Parent CG shown in figures labeled '00:33' and '00:58' were also the same. Figure 32 shows that radar reflectivity characteristics at these two moments were similar, therefore, vertical structure at 00:56 (TRMM passed the storm at this time) could present the storm structure at 00:33 to a large extent. Its vertical slices will be shown in the next section.

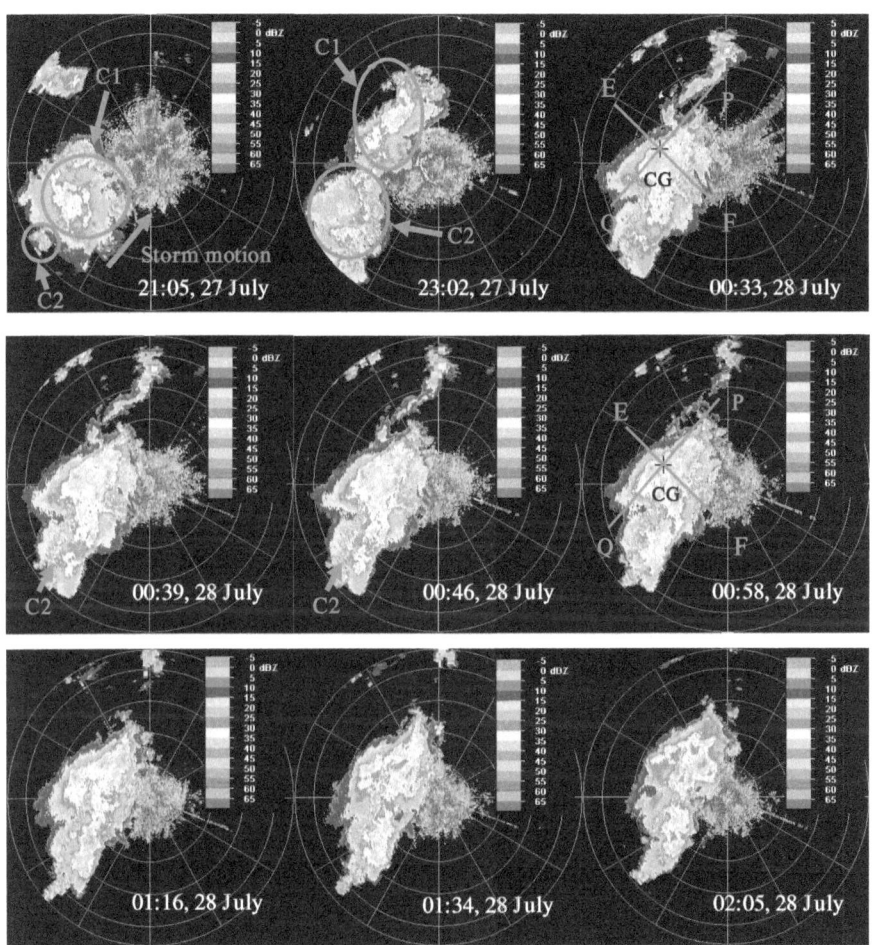

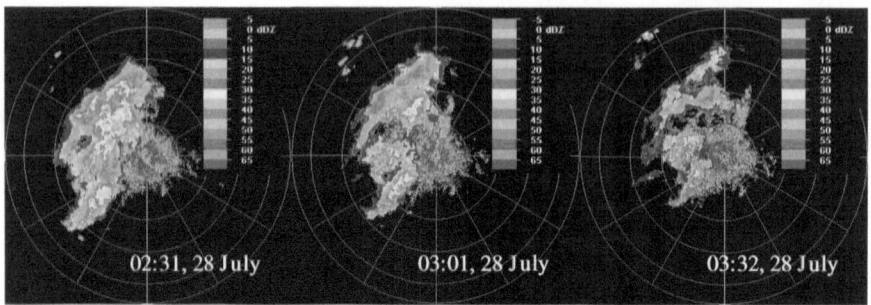

Figure 32. Radar reflectivity of the storm on 27-28 July obtained from the Doppler radar data. The time and storm motion are also shown in the figure ('+' in the figure labeled parental CG). Parent CG in '00:58' and '00:33' images is the same

Figure 33 shows that the 29-30 July storm occurred at 19:00 with maximum composite radar reflectivity of 40-45 dBZ. After an hour, the storm developed and several convective centers (maximum radar reflectivity of 50 dBZ) could be seen at 20:05 (as shown in Figure 33). At 20:29, a comparatively large center (labeled 'C' in the figure) occurred, but weakened rapidly (as shown in the figure labeled '21:05'). This convective center was weaker than that in storms on 1-2 August and 27-28 July. At 22:05, the storm area became larger compared with that at 21:05. Doppler radar images show that comparatively large stratiform regions could be seen at 02:00 30th July. The stratiform region in this storm was larger than that in 27-28 July storm. Compared with the two sprites-producing storms, no very strong convective center occurred during this storm evolution. Very strong convective centers in the two-sprites producing storms provide favorable condition for development of large stratiform regions and for sprites.

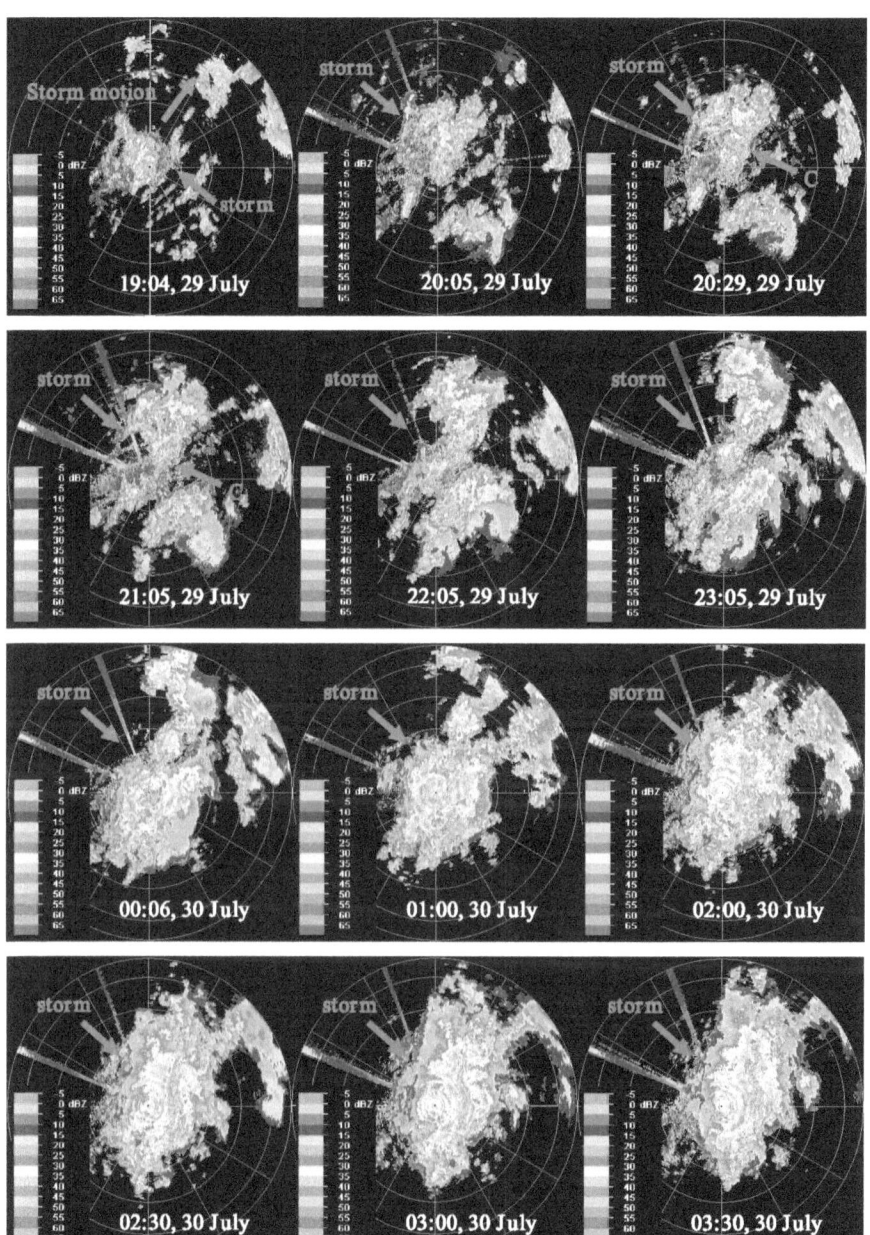

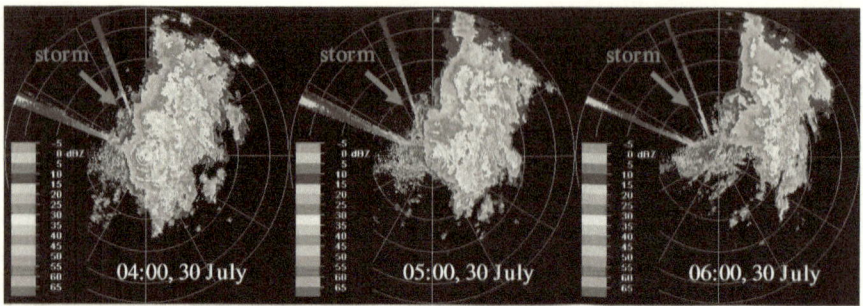

Figure 33. Radar reflectivity of the storm on 29-30 July obtained from the Doppler radar data. The time and storm motion are also shown in the figure

It is shown from above analysis that common property of the three storms was that each thunderstorm consisted of multiple convective cells. With the dissipation of previous cell, the newly formed one strengthened. The difference between three thunderstorms is that very strong convective centers occurred during sprites-producing storms, but not in the non-sprites producing storm. Large stratiform regions occurred during sprites-producing storms, and the parent CGs located in these stratiform regions with radar reflectivity of 15-35 dBZ, similar to previous studies (e.g. Lyons, 1996; Lyons et al., 2003). Large stratiform regions also occurrred in the non-sprites-producing storm, but no sprites were recorded. Microphysical structures of the three storms will be given in the next section.

3.4 Thunderstorm microphysical structure

The three storms were fortunately scanned by TRMM, and valuable data were thereby acquired. Since the swath width of PR (247 km) is much smaller than that of TMI (878 km), only parts of the storms were scanned by PR, the PR data will not be used in this paper (PR scanned the whole storm on 1-2 August, and its information could be found in Yang et al. (2011)). The TRMM data used in this paper is TMI

(2A12). As mentioned in previous section, some parent CGs in the two sprites-producing storms have been found. Characteristics in regions where parent CG located were analyzed by using TRMM data. Vertical slices along lines GH, MN (shown in Figure 34, the locations of lines GH and MN are the same as in Figure 31) and EF, PQ (shown in Figure 35, the locations of lines EF and PQ are the same as in Figure 32) were made. Lines GH and EF are perpendicular to the storm movement direction. Lines MN and PQ are parallel to the storm movement direction. The results show that precipitation ice, cloud ice and cloud water along these lines were uniform, no special feature related to sprites. The maximum values of precipitation ice, cloud ice and cloud water contents along these lines are listed in Table 2. Table 2 shows that the maximum values of precipitation ice, cloud ice and cloud water in strong radar reflectivity regions were larger than that in weak radar reflectivity regions. In addition, cloud ice were very small in these regions, most of the particles along these lines were precipitation ice.

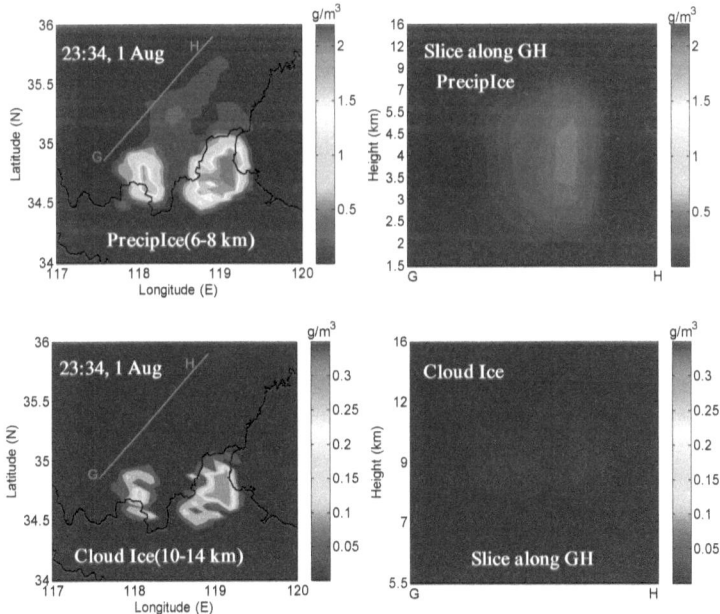

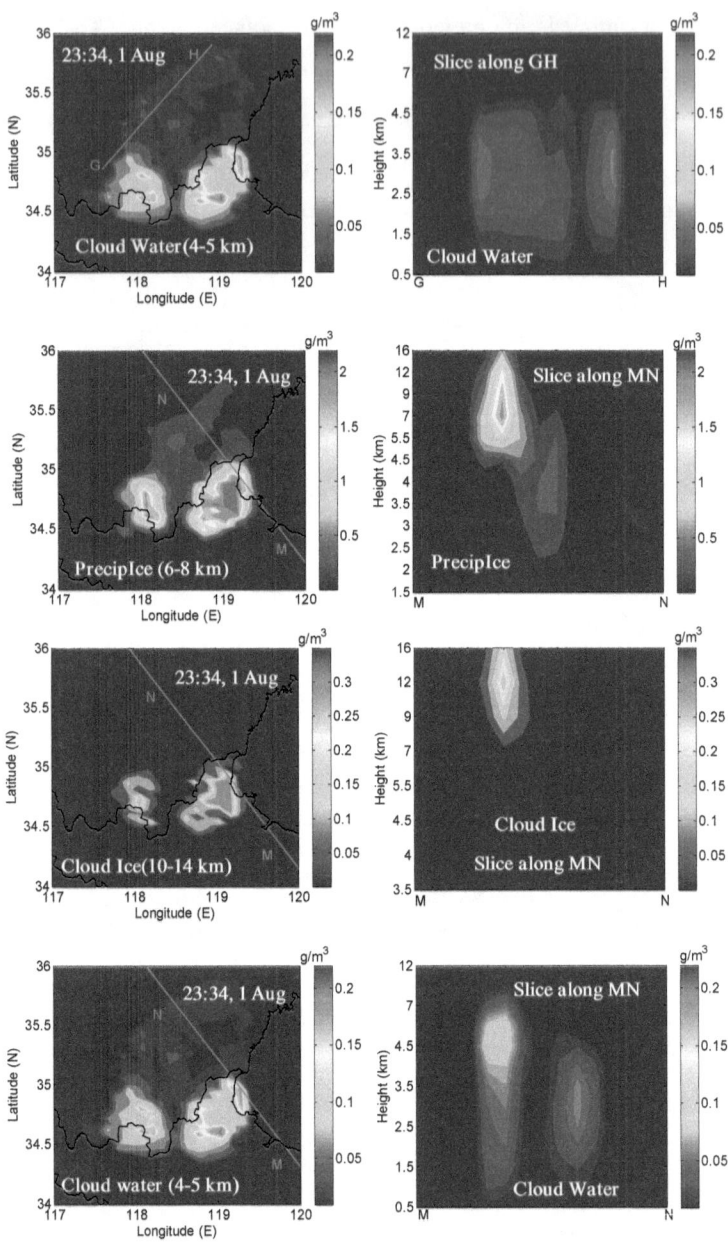

Figure 34. Vertical cross sections of precipitation ice, cloud ice and cloud water along lines GH and MN in 1-2 August storm. The locations of lines GH and MN are the

same as in Figure 31

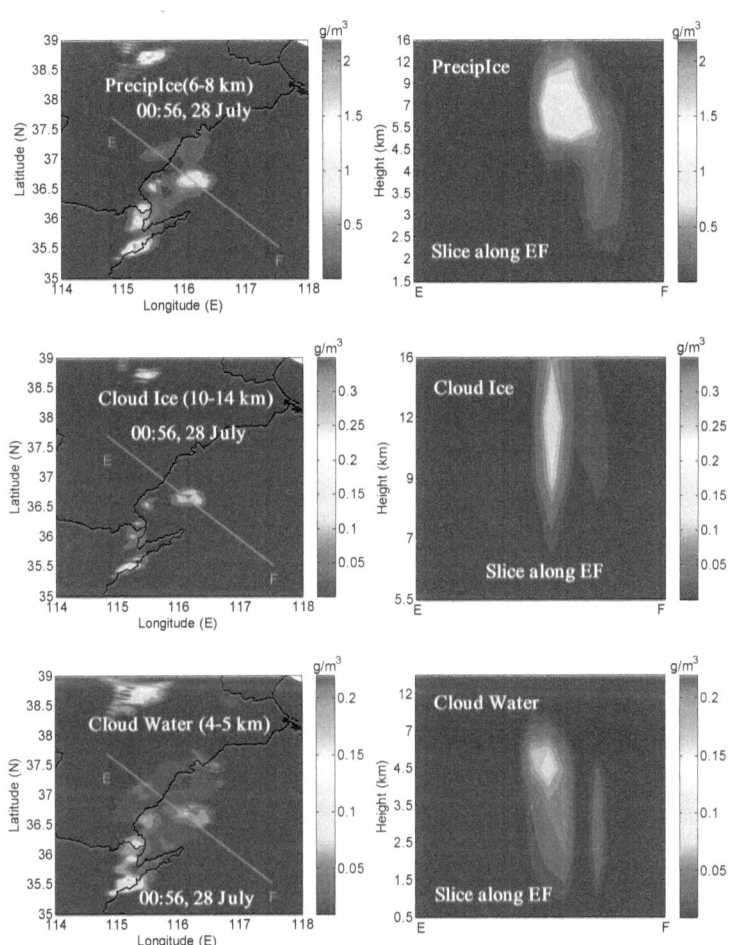

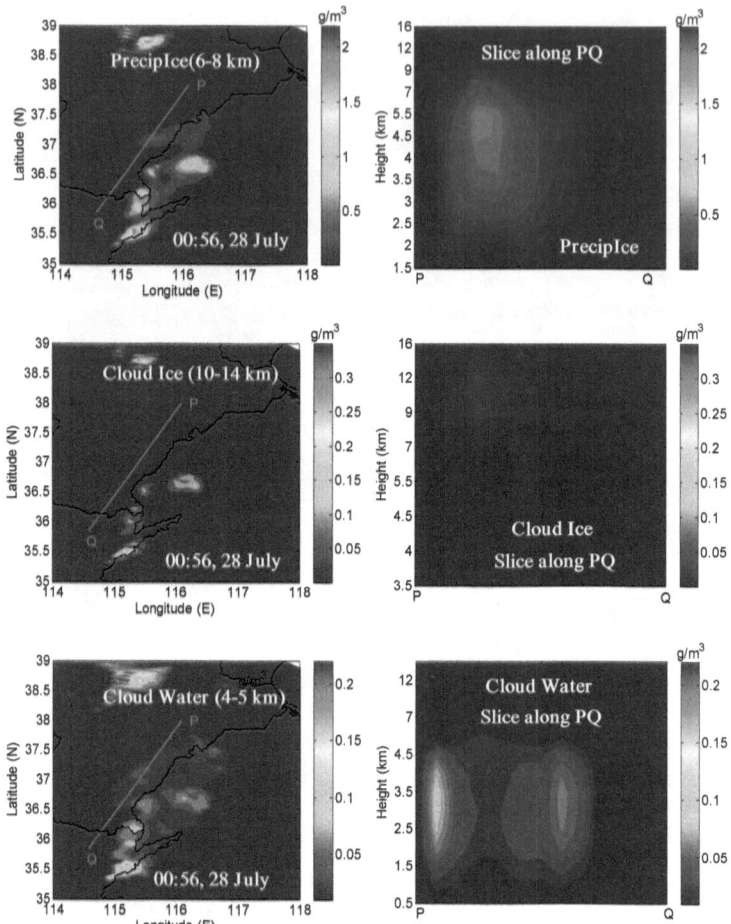

Figure 35. Vertical cross sections of precipitation ice, cloud ice and cloud water along lines EF and PQ in 27-28 July storm. The locations of lines EF and PQ are the same as in Figure 32

Table 2. Microphysical characteristics of the two sprites-producing storms

	1-2 August	27-28 July
Time when TRMM	23:34 1st August	00:56 28th July

passing by

	Along line GH	Along line MN	Along line EF	Along line PQ
Precipitation ice (g/m^3)	0.544	1.76	1.152	0.59
Cloud ice (g/m^3)	0.016	0.276	0.162	0.026
Cloud water (g/m^3)	0.062	0.11	0.094	0.091

From above analysis, it is known that vertical structures along lines through parent CG location showed no special characteristics related to sprites. Since sprites can be tens of kilometers away from its parent CGs (Sâo Sabbasa et al., 2003; Soula et al., 2010), microphysical structure in other regions of the storm may also play important roles in determining sprites. Sprites are known to be more related with positive CGs located in large stratiform regions, so vertical slices along line CD in stratiform region and along line AB in convective region were analyzed and shown in Figures 36-38. The precipitation ice, cloud ice and cloud water content in convective and weak regions were listed in Table 3. The results show that precipitation ice, cloud ice and cloud water in convective region in 1-2 August storm were larger than that in the other two storms, but it seemed contrary in weak regions. This result may to some extent indicate that convection in the other two storms were not very strong. The cloud water in convection region in 29-30 July storm has the maximum value below 2.5 km, which was similar to that in 1-2 August storm. It should also be mentioned that these values may change if the location of lines changed.

Figures 36-38 also show that most precipitation ice located at 6-8 km for the three storms. The three storms have similar distributions with most cloud ice at 10-14 km and cloud water at 4-5 km (some figures for 1-2 August storm were adopted from Yang et al. (2011) and were shown here for convenience). Although the microphysical structure in the non-sprites producing storm was similar to that of the sprites-producing storm, no sprites were recorded. The TRMM and Doppler radar figures almost show the same time and geographic area for the three storms, so the Doppler radar images were also shown in Figure 36. It is shown that regions with radar reflectivity of 15-30 dBZ in 1-2 August storm was concentrated in the storm rear regions (the storm motion was also shown in the figure), this storm produced the largest number of sprites. By comparing the radar reflectivity at 00:56 28th July and 00:42 30th July, it indicates that comparatively large weak regions (15-20 dBZ) could be found in the rear in 28th July storm and one sprite was recorded over this storm. Apparently, the death of parts of the convective region can introduce favorable conditions in charge distribution or triggering of sprites-producing positive discharges. Or perhaps the death of the convective region and growth of stratiform region are caused by a more slantwise front-to-rear flow in the MCS, which could result from a change in the balance between inflow winds and the outflow. Stratiform regions also occurred in the 29-30 July storm, but no sprites were recorded during the whole night. The echo top mainly located between 9 and 12 km with the maximum of 17 km at 23:35 in 1-2 August storm (see Table 3), which is a little higher than result (16.7 km) reported by Sâo Sabbasa et al. (2010). The echo top in 27-28 July storm mainly situated between 6 and 9 km with maximum value of 14 km at 00:56, indicating the convection was not very strong. For storm on 29-30 July at 00:42 the maximum echo top was 14 km with most at 6-9 km, also indicating the convection was not very strong. Comparative analysis showed that the 1-2 August storm was the strongest and it was most effective in creating a charged stratiform region for sprites. The meteorological characteristics, microphysical structure and radar reflectivity pattern of the non-sprites-producing storm on 29-30 July was similar to the most sprites-producing storm on 1-2 August, but no sprites were recorded. Since two orbit data could be used for 29-30 July storm, the microphysical characteristics of this storm were shown in Figures 39-41. Although these figures showed no special features, anyway, they provide more information of this non-sprites-producing storm and were shown here.

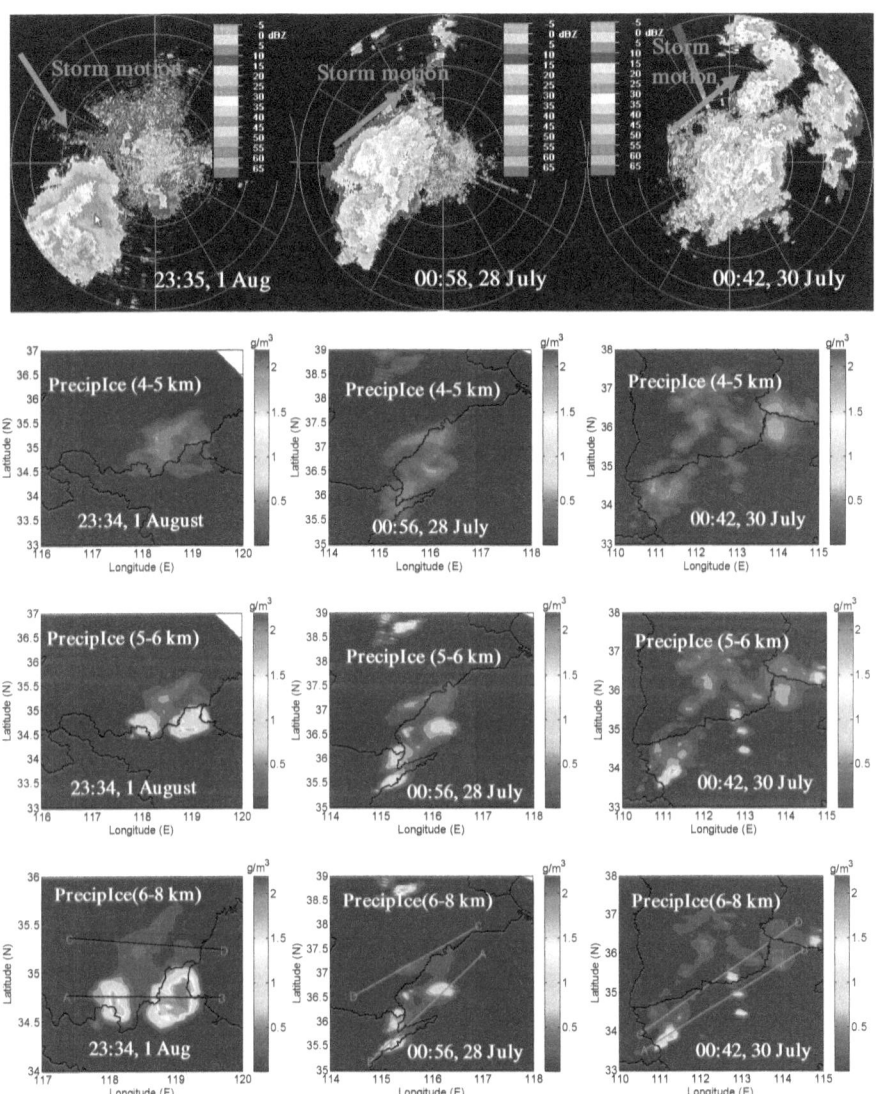

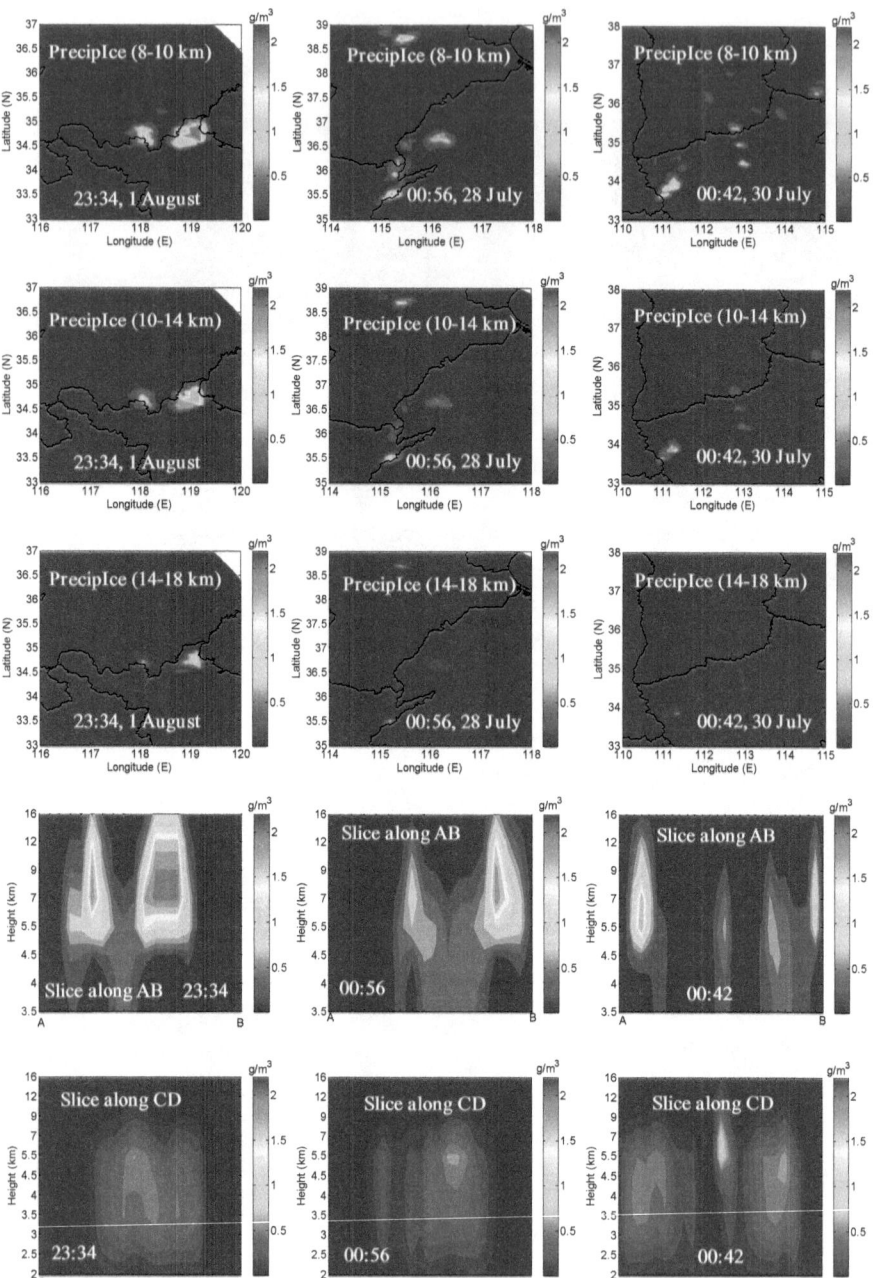

Figure 36. Radar reflectivity in the three storms obtained from Doppler radar.

Precipitation ice at different levels and slices of precipitation ice content in the three storms along lines AB and CD as shown in the fourth row in the figure. The time is shown in the figure

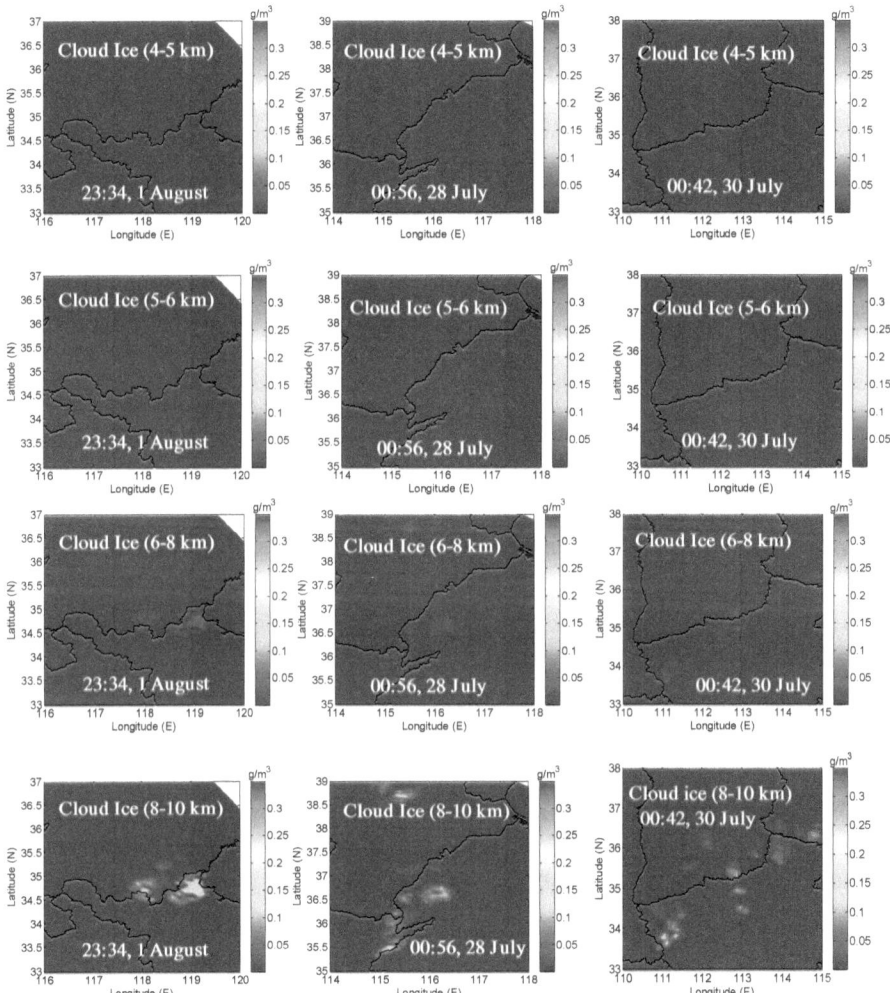

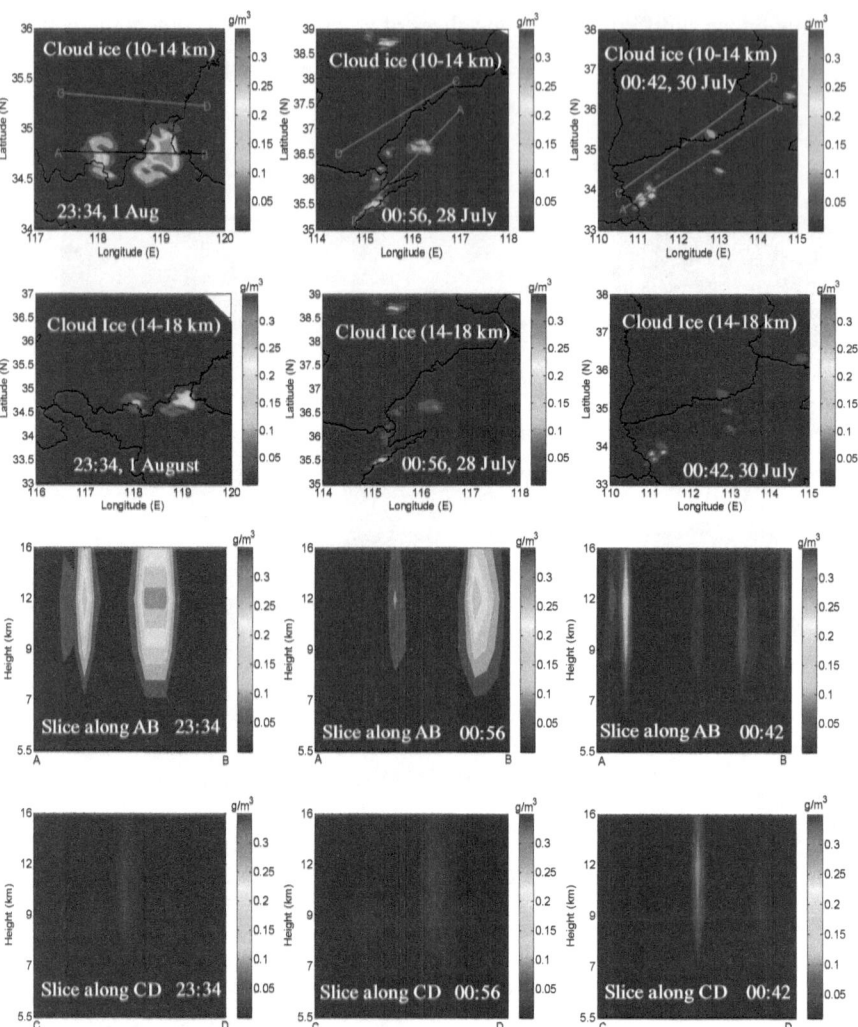

Figure 37. Cloud ice content at different levels in the three storms. Slices of cloud ice content along lines AB and CD as shown in the fifth row in the figure. The time is also shown in the figure

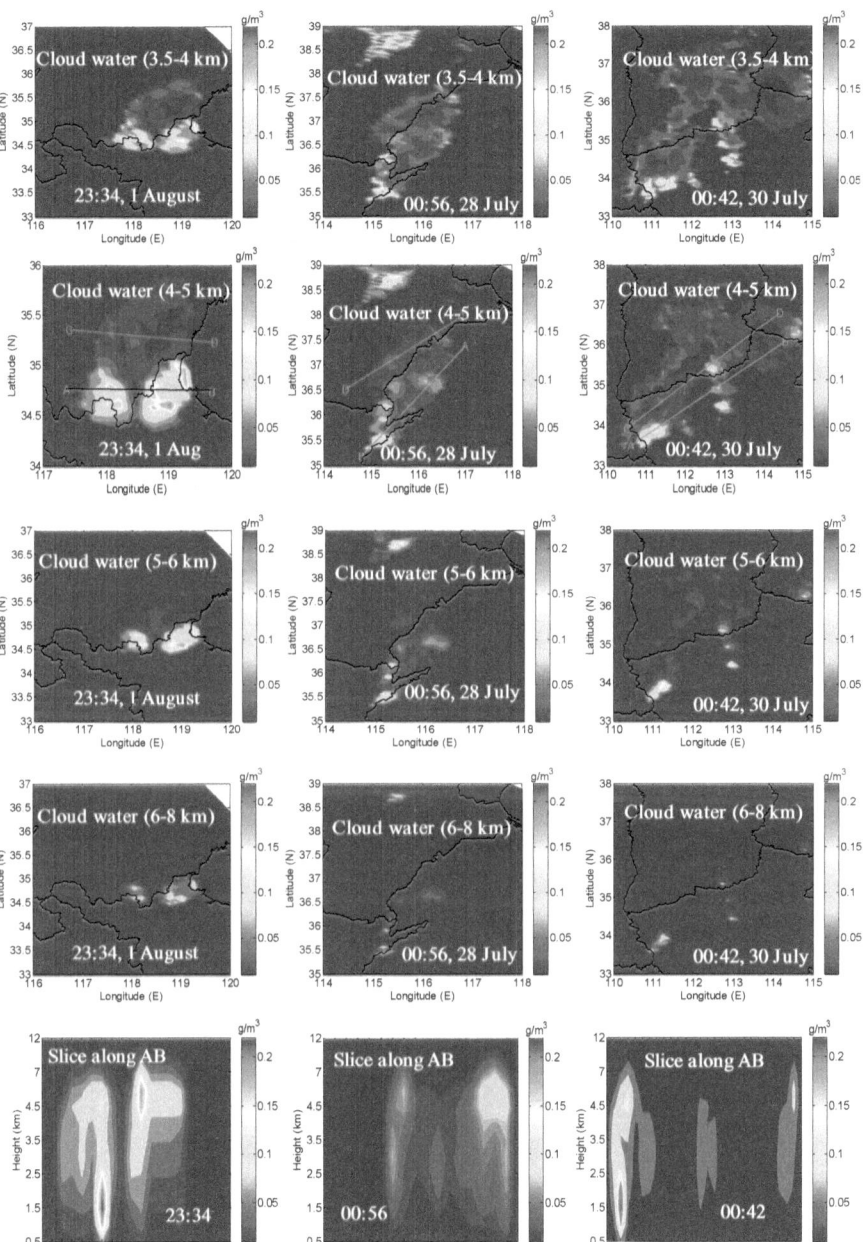

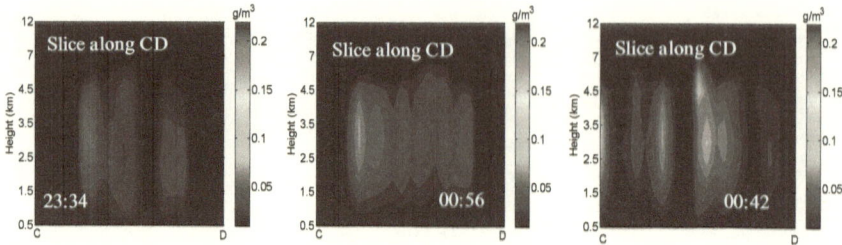

Figure 38. Cloud water content at different levels in the three storms. Slices of cloud water content along lines AB and CD as shown in the second row of the figure. The time is also shown in the figure

Table 3. Microphysical characteristics of the three storms

	1-2 August		27-28 July		29-30 July	
Time when TRMM passing by	23:34 1st August		00:56 28th July		00:42 30th July	
Recorded TLEs	16		1		0	
Maximum echo top	17 km		14 km		14 km	
30-35 dBZ echo top	9-12 km		6-9 km		6-9 km	
	Convection region	Weak region	Convection region	Weak region	Convection region	Weak region
Precipitation ice (g/m^3)	2.05	0.609	1.964	0.633	1.81	0.903
Cloud ice (g/m^3)	0.326	0.032	0.301	0.032	0.161	0.111
Cloud water (g/m^3)	0.232	0.057	0.119	0.073	0.261	0.095

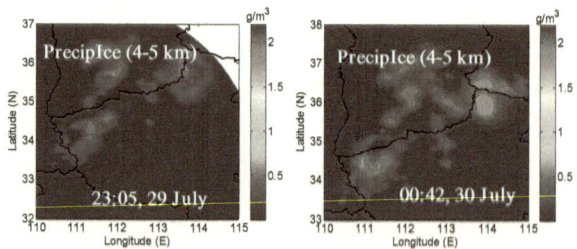

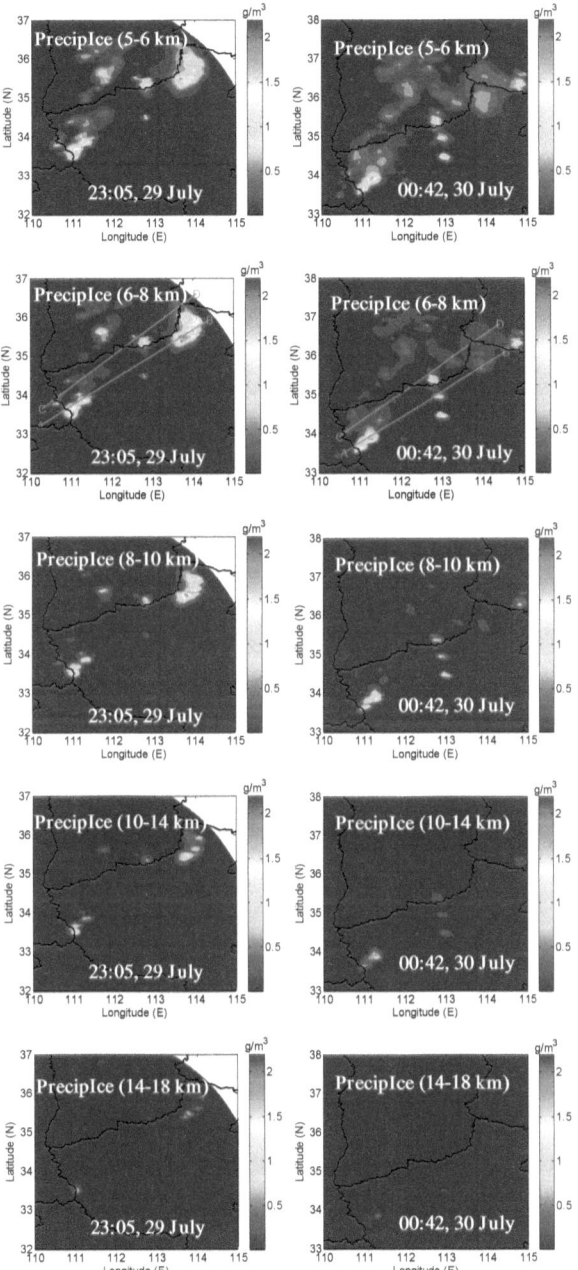

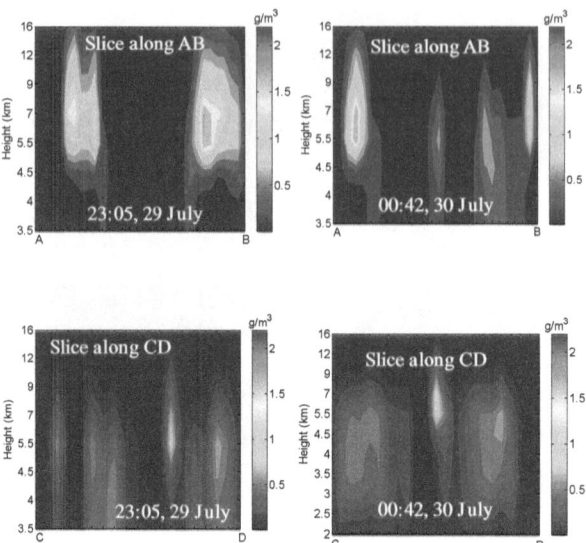

Figure 39. Precipitation ice content at different levels and vertical structures along lines AB and CD at two moments obtained from TRMM 2A12 data. The time is also shown in the figure

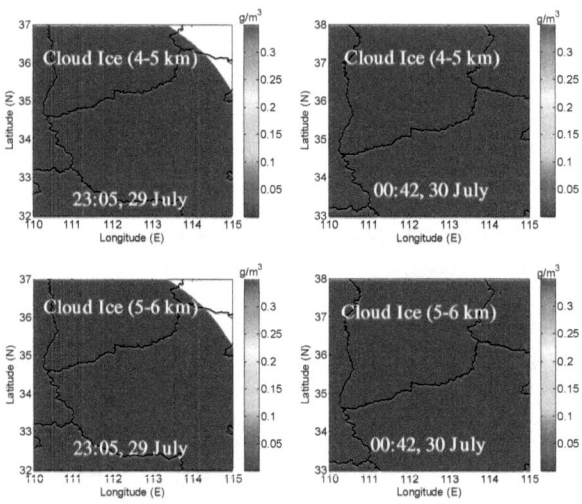

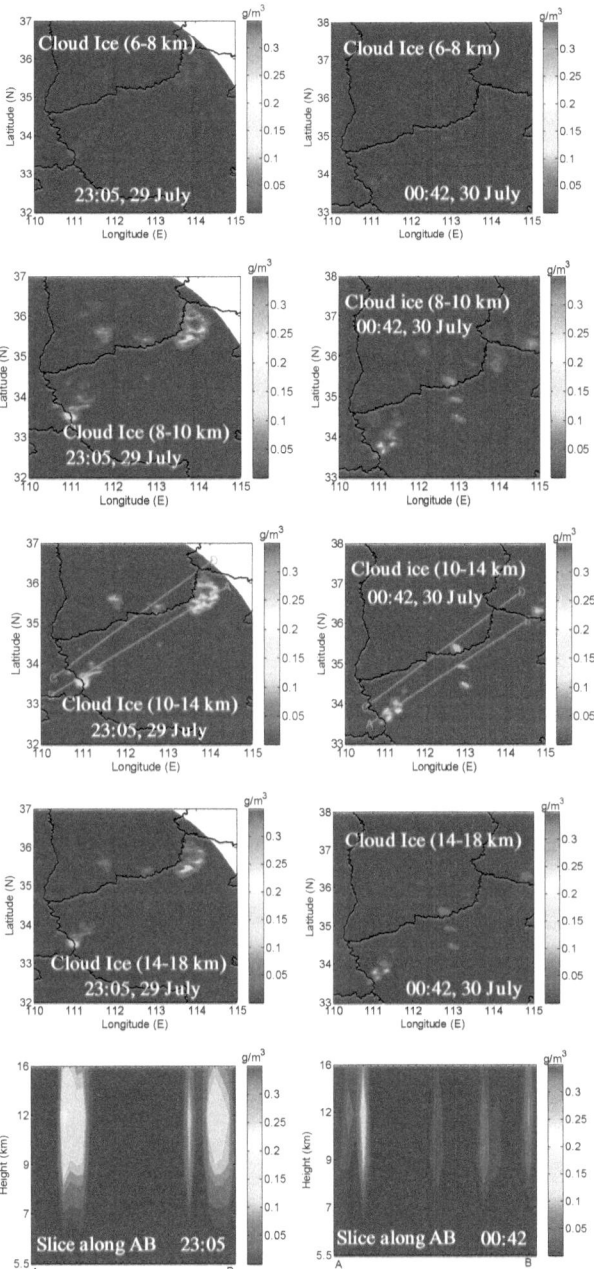

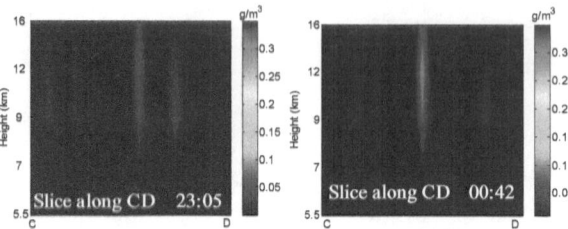

Figure 40. Cloud ice content at different levels and vertical structures along lines AB and CD at two moments obtained from TRMM 2A12 data. The time is also shown in the figure

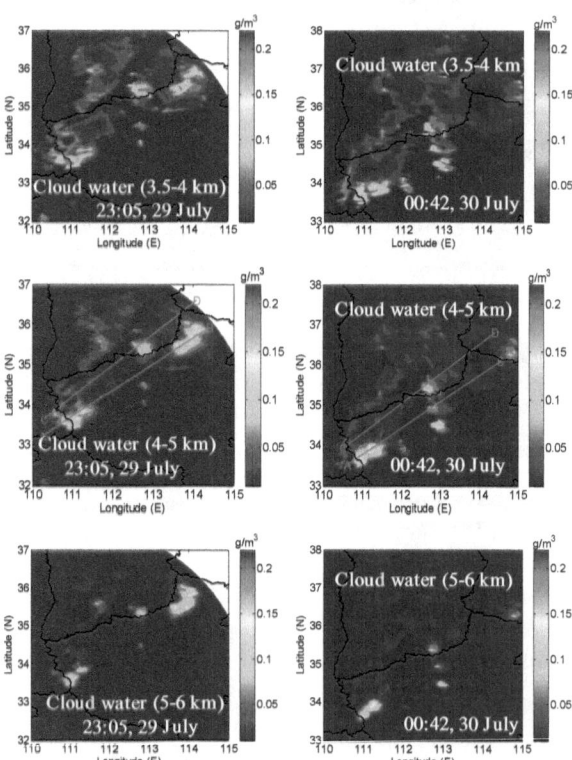

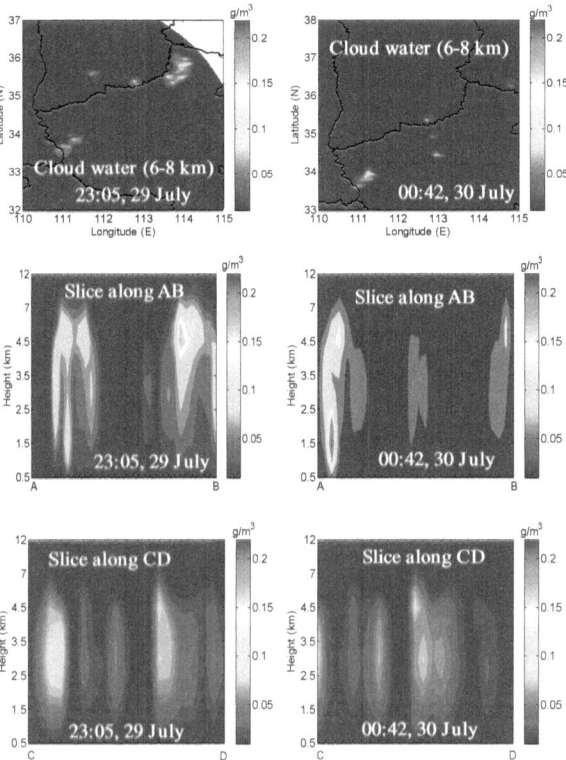

Figure 41. Cloud water content at different levels and vertical structures along lines AB and CD at two moments obtained from TRMM 2A12 data. The time is also shown in the figure

3.5 Characteristics of thunderstorm lightning activities

Since lightning flash produces sprites, are there any difference in lightning activity between sprites-producing and non-sprites producing storms? Characteristics of lightning activity will be analyzed in this section. The lightning data are provided by lightning location network which has a detection efficiency around 90% (Feng et al., 2007). The sprites were not triangulated and it is assumed that sprites centered above the causative +CGs. Figure 42 shows the spatial distribution of negative CGs, positive CGs and parent CGs, it was found that the 29-30 July storm produced the largest number of flashes. The parent CGs clustered in a comparatively small region in 1-2 August storm, this region may have favorable condition for sprites.

Figure 43 shows the evolutions of CG lightning flash rate per 5 minutes during the three thunderstorms. There was an obvious jump in CG flash rate during the storm rapidly developing stage and negative CGs predominated during this stage. Positive CG lightning reached a peak with flash rate of 4 fl/5 min before 21:00. The positive CGs rate reached the largest value of 6 fl/5 min around time 02:00. During the sprite time period (00:00-02:00, 2nd August), negative CG flash rate declined rapidly while positive CG flash rate increased, which lead to a high proportion of +CGs, in good agreement with previous study (Soula et al., 2009).Negative CG lightning became active again after the sprite period (00:00-02:00, 2nd August). From the evolution of Doppler radar images, it was found that the storm began to weaken during the movement towards the southeast since 22:30, and simultaneously the radar reflectivity went down from 50 dBZ at 22:30 to 35 dBZ at 00:00. But forced by outflow from the thunderstorm, weak convective cells were continuously initiated and subsequently merged into the old storm, causing convection strengthening. This was the reason why negative CG lightning became active again after the sprite period.

The overall evolution curves of the CGs in thunderstorm on 27-28 July is obviously different from that in storm on 1-2 August. The positive CG flash rate in 27-28 July storm reached the maximum value of 16 fl/5 min at about 18:00 (radar reflectivity of 50 dBZ) during the storm developing stage, but positive CG flash rate in storm on 1-2 August reached its maximum value during the storm dissipating stage, indicating that the lightning activities in the two thunderstorms is very different. At about 22:05 27th July, the negative CG flash rate was about 32 fl/5 min, while positive CG was 1 fl/ 5 min. The sprite occurred at 00:36:39 28th July, the positive CG rate was 6 fl/5 min and negative CG rate was 9 fl/ 5 min at this time. The evolution curve of CG flash rate is erratic in 27-28 July storm compared with that in 1-2 August storm, but it is confirmed that only lightning flashes that related to the sprite-producing storm cell were included. Generally speaking, positive CG rate in 27-28 July storm was larger than that in 1-2 August storm. Since most observed sprites are associated with positive CGs, maybe more sprites could be expected in 27-28 July storm than that in 1-2 August storm. But the results were not as expected. It should be noted that there was fog in the camera's field view at the late stage of 27-28 July storm, so whether some positive CGs produced sprites or not is not known from the single optical observation (Multi-station obervations will be considered in the future). The lightning activity in 29-30 July storm showed similar characteristics as 1-2 August storm, with negative CGs dominate during the storm developing stage, positive CGs increased and negative CGs decreased (corresponded with decrease of convective intensity) during the storm dissipating stage. Although characteristics of CG flash rate evolutions in 29-30 July storm are similar to 1-2 August storm, no sprites were recorded over 29-30 July storm.

It is worth noting that locations of parental CGs, occurred at 01:12:31, 01:10:42 and 01:18:27 2nd August, were very close. The shortest distance between the two of them (01:12:31 and 01:10:42, 01:18:27 and 01:10:42) was about 13 km, and the largest (01:18:27 and 01:12:31) was about 26 km. Compared with distance (about 300 km) between the storm and the observation site, the 13 and 26 km were very small, and could be considered almost in the same location in the storm, which was similar to results obtained in Northern America (Winckler, 1995). Apparently, this region in the cloud has favorable condition for sprites. The storm electrical structure and some unknown factors may play important roles. Further studies are needed.

In this study, peak current of parental CGs less than 100 kA accounted for 62.5%, larger than 100 kA only 37.5%, in good agreement with Sâo-Sabbas et al. (2003) and Winckler et al. (1996). Figure 44 shows the evolution of positive CGs peak current in the three thunderstorms. Table 4 and Figure 45 show the positive CG statistical characteristics. The results show that average value (average value was 112 kA, geometric mean was 82 kA) of parent CGs in 1-2 August storm was larger than that of all positive CGs (average value of all positive CGs was 82 kA, and geometric mean was 71 kA), consistent with results reported by Soula et al. (2009) in which parent CG peak current was twice the average value of positive CGs. During the time centered at 01:20 in 1-2 August the sprite production rate was about one every 9 minutes and three times larger than that during the time period centered at 22:00. It is also shown that the number of positive CGs at 01:20 was larger than that at 22:00. The parental CG peak current for sprtie at 00:36:39 in 27-28 July storm was 196 kA, a relatively large value. Both the average value and geometric mean of positive CGs in 27-28 July storm were larger than that in 1-2 August storm, and its maximum value (427 kA) was also larger. The average value of positive CG peak current in 29-30 July storm was 69 kA and geometric mean was 53 kA, smaller than that in 1-2 August storm and 27-28 July storm. These discussions show that parent CG peak current was not very large compared with the maximum value of positve CG peak current in the storm, but the average value of positive CG peak current in sprites-producing storms were larger than that in the non-sprites-producing one. Figure 45 shows that positive CG peak current in 1-2 August storm has a maximum around 60-70 kA, and the 29-30 July storm has a maximum around 30-40 kA. It is interesting that the 27-28 July storm has maximum value around 40-120 kA, which is a very wide range compared with storms on 1-2 August and 27-28 July. Percent of positive CGs in 1-2 August and 27-28 July storms was also larger than that in 29-30 July storm, but is lower than results (40%) obtained by Soula et al. (2010). It should be noted that positive CG strokes with peak current less than 10 kA are likely to be intracloud flashes that have been filtered out in this paper, and no filtering was made in Soula et al. (2010). Since this paper is a case study, and more cases are needed to test the validity and universality of these results.

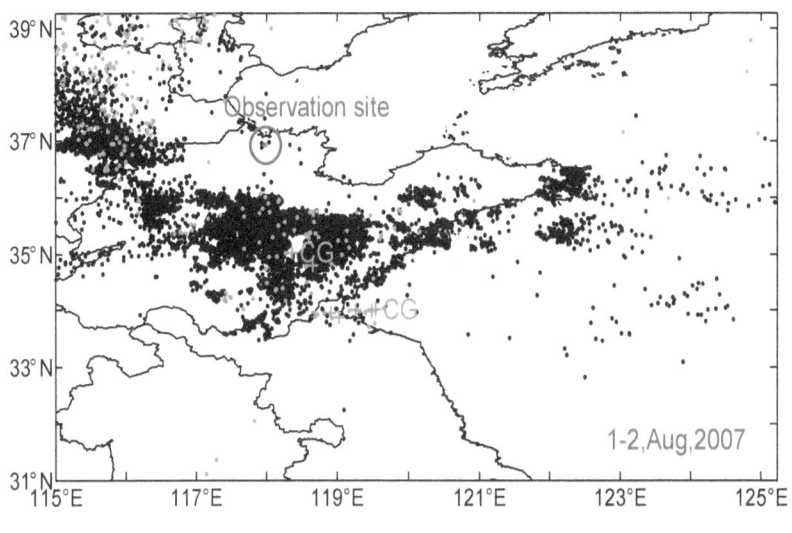

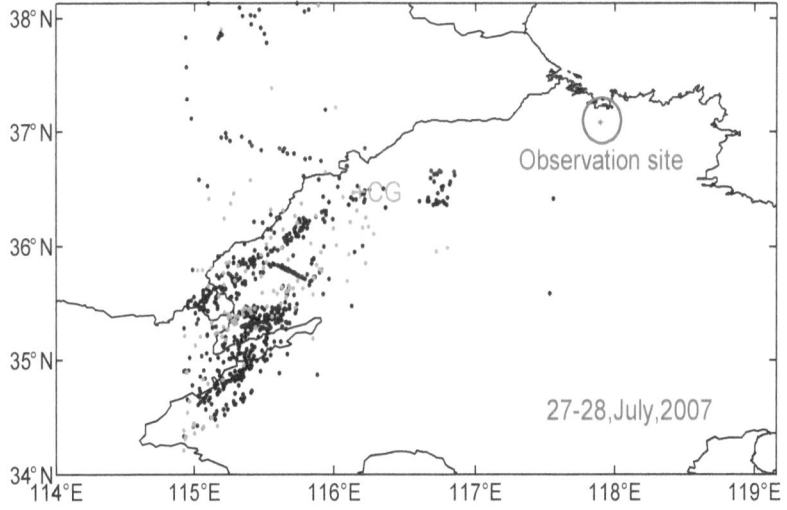

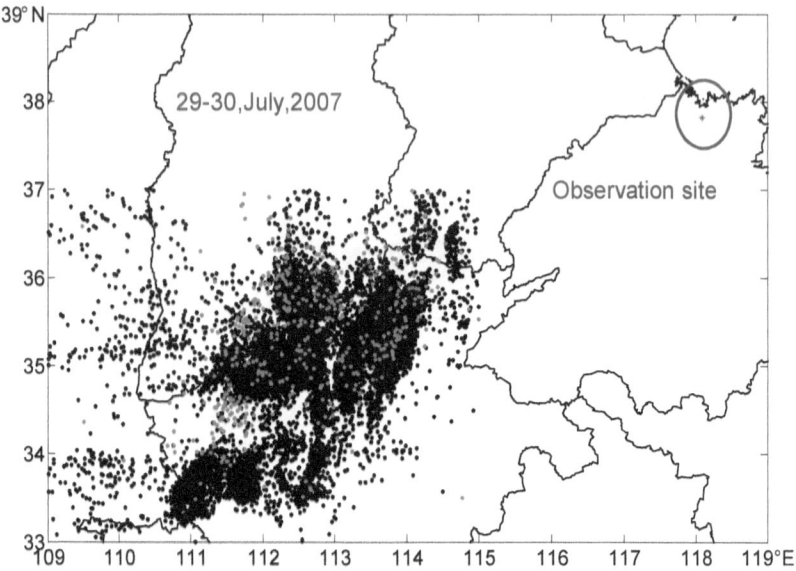

Figure 42. Distribution of total CG lightning during the whole process in the three thunderstorms. The red '.' and black '.' stand for positive and negative CG lightning, respectively. The red symbol '+' stand for parental CGs. The circled blue '*' stands for the observation site

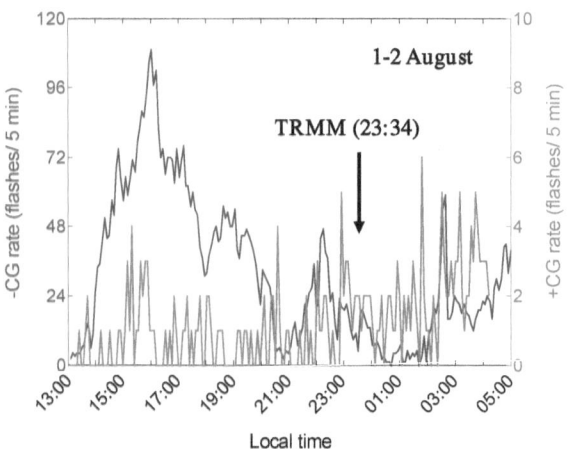

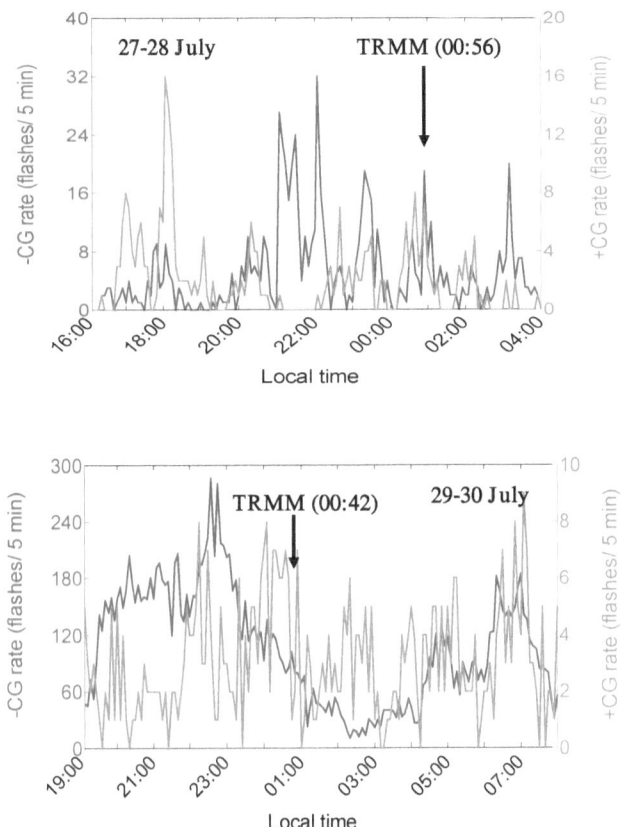

Figure 43. Evolutions of negative and positive CG lightning flash rate per 5 minutes during the three thunderstorms. The black arrow indicate the moment when TRMM passing by

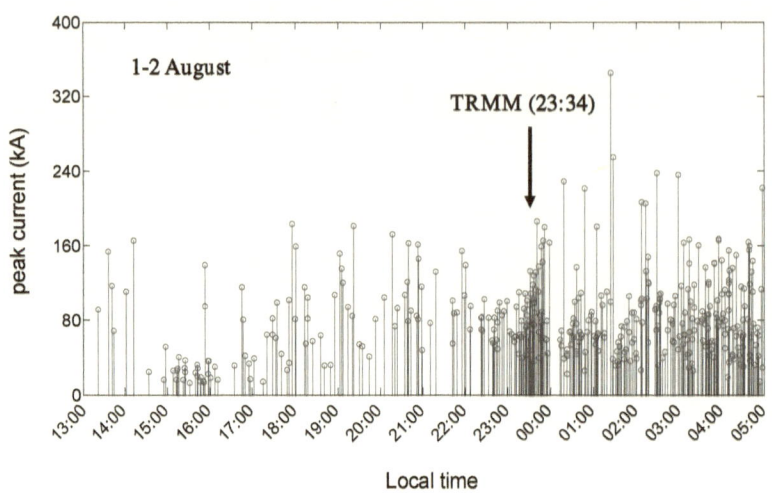
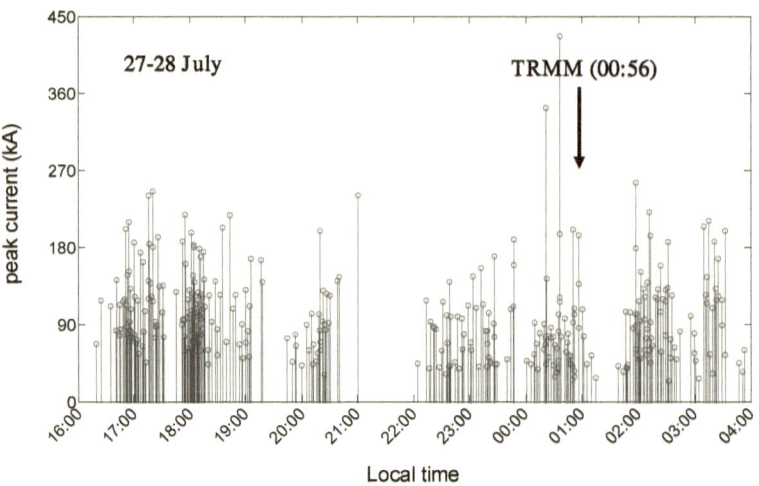

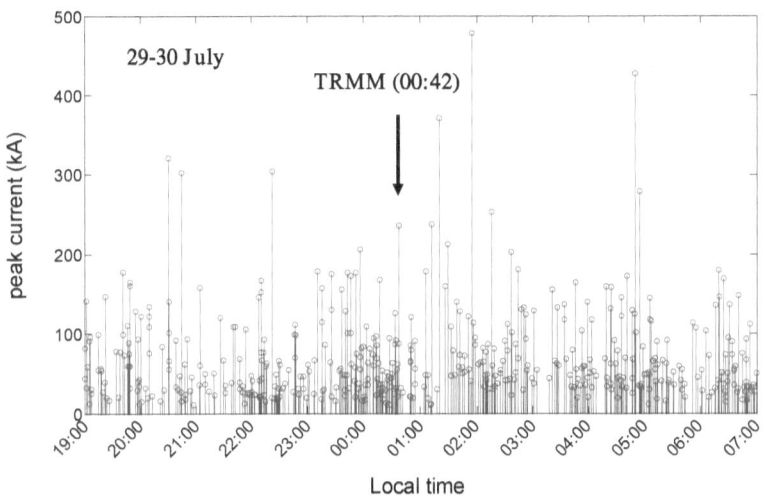

Figure 44. Evolutions of positive CGs in the three thunderstorms. The black arrow indicate the moment when TRMM passing by

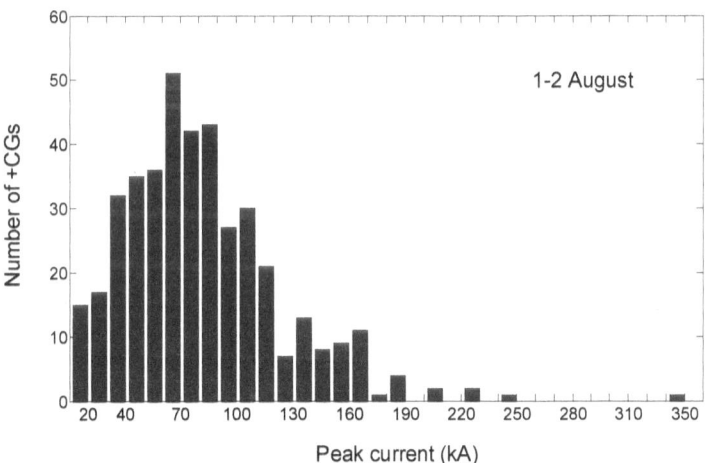

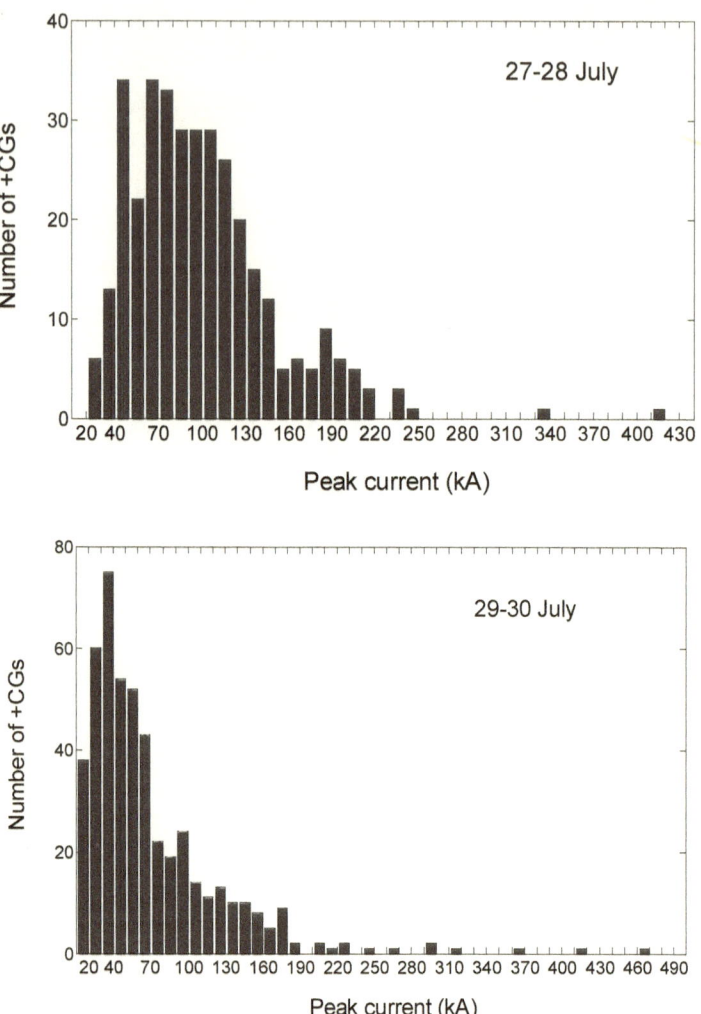

Figure 45. Statistical distribution of positive CG peak current in the three thunderstorms

Table 4. Characteristics of positive CG peak current in the three storms

Peak current (kA)	1-2 August	27-28 July	29-30 July	Parent CGs in 1-2 August

				storm
>250	0.5%	0.8%	1.7%	0
>200	2.2%	4.2%	2.7%	28%
>150	8.4%	13.1%	7.7%	28%
>100	27.5%	42.5%	19.7%	28%
>50	76.2%	84.6%	52.8%	71%
Percent of positive CGs	2.2%	24.4%	0.7%	
Average Value	82 kA	99 kA	69 kA	112 kA
Geometric Mean	71 kA	88 kA	53 kA	82 kA
Standard Deviation	44 kA	51 kA	56 kA	84 kA

4. Summary

This study describes the meteorological backgrounds for three nights of thunderstorms in eastern China, using multiple data from Doppler radar, MTSAT infrared images, lightning location network, TRMM satellite, NCEP and radiosonde. Two nights of storms produced a total of seventeen sprites. And no sprites were recorded over the third storm. Considering the most sprites-producing storm on 1-2 August, the flash rate graph provides additional confirmation that major sprite activity are introduced by strong decrease in negative CG flash rate. The lightning behavior of non-sprites-producing storm was similar to that in 1-2 August storm, but no sprites were recorded during the whole night. The flash rate evolution in 27-28 July storm seems erratic compared with that in 1-2 August, but one sprite was observed. All of the sprites in this study were produced by positive CGs that remove charges from stratiform region, which also reinforces previous studies (Lyons, 1996; Lyons et al., 2003; Soula et al., 2009, 2010). The average positive CG peak current in sprites-producing storms were larger than that in the non-sprites-producing one.

Microphysical structures and parameters were analyzed by using TRMM for the three storms. The storm vertical structures along lines through parent CG location showed no special characteristics related to sprites. Convection in the most sprites producing storm on 1-2 August was stronger than that in the other two storms, which was effective in creating a charged stratiform region for sprites. The precipitation ice, cloud ice and cloud water in convection region in 1-2 August storm were larger than that in the other two storms, but it seemed contrary in weak regions. The three storms have similar distributions with most precipitation ice located at 6-8 km, cloud ice at 10-14 km and cloud water at 4-5 km.

Acknowledgement: This research was supported jointly by National Natural Science Foundation of China (41374153, 8143038), the Youth Innovation Promotion Association of Chinese Academy of Sciences and the high resolution Earth observation funds for young scientists (Grant NO. GFZX04060103-7-11). All the authors would like to thank the Kochi University, Wyoming University, NOAA/OAR/ESRL PSD, NASA/Goddard Earth Sciences/Data and information Services Center, NCEP, TRMM data.

References:

Adachi, T., H. Y. Fukunishi, and Y. Takahashi, Electric field transition between the diffuse and streamer regions of sprites estimated from ISUAL/array photometer measurements, Geophy. Res. Lett., 2006, 33(L17803), doi: 10.1029/2006GL026495.

Barrington-Leigh C P, Inan U S, Elves triggered by positive and negative lightning discharges, Geophys. Res. Lett., 1999, 26(6), 683-686.

Barrington-Leigh C P, Inan U S, Elves triggered by positive and negative lightning discharges, Geophys. Res. Lett., 1999a, 26(6), 683-686.

Barrington-Leigh, C. P., Inan, U. S., Stanley, M., Cummer, S. A., 1999b, Sprites triggered by negative lightning discharges. Geophy. Res. Lett., 26 (24), 3605-3608.

Barrington-Leigh, C. P., Inan, U. S., Stanley, M., 2001, Identification of sprites and elves with intensified video and broadband array photometry. J. Geophys. Res., 106 (A2), 1741-1750.

Bell, T. F., Reising, S. C., Inan, U. S., 1998, Intense continuing currents following positive cloud-to-ground lightning associated with sprites. Geophys. Res. Lett., 25(8), 1285-1288.

Boccippio, D. J., Williams, E. R., Heckman, S. J., Lyons, W. A., Baker, I. T., Boldi, R., 1995, Sprites, ELF transients, and positive ground strokes. Science, 269 (5227), 1088-1091.

Boeck, W. L., O. H. Vaughan, R. J. Blakeslee, B. Vonnegut, and M. Brook (1992), Lightning induced brightening in the airglow layer, Geophys. Res. Lett., 19, 99-102.

Boeck, W.L., O. H. Vaughan, R. J. Blakeslee, B. Vonnegut, and M. Brook (1998), The role of the space shuttle videotapes in the discovery of sprites, jets and elves, J. Atmos. Sol. Terr. Phys., 60, 669-677.

Chang, S. C., C. L. Kuo*, L. J. Lee, A. B. Chen, H. T. Su, R. R. Hsu, H. U. Frey, S. B. Mende, Y. Takahashi and L. C. Lee, 2010, ISUAL far-ultraviolet events, elves, and lightning current, J. Geophys. Res., Volume 115, A00E46.

Chen, A. B., Kuo, C. L., Lee, Y. J., Su, H. T., Hsu, R. R., Chern, J. L., Frey, H. U., Mende, S. B., Takahashi, Y., Fukunishi, H., Chang, Y. S., Liu, T. Y., Lee, L. C., 2008, Global distributions and occurrence rates of transient luminous events. J. Geophys. Res., 113(A08306), doi:10.1029/2008JA013101.

Chen, A. B., H. T. Su, and R. R. Hsu, 2014, Energetics and geographic distribution of elve-producing discharges, Journal of Geophysical Research: Space Physics, Volume 119, 1381-1391.

Chou, J. K., L. Y. Tsai, C. L. Kuo, Y. J. Lee, C. M. Chen, A. B. Chen, H. T. Su, R. R. Hsu*, P. L. Chang and L. C. Lee (2011), Optical emissions and behaviors of the blue starters, blue jets, and gigantic jets observed in the Taiwan transient luminous event ground campaign, Journal of Geophysical Research: Space Physics, Volume 116, A07301, doi:10.1029/2010JA016162.

Chou J K, Kuo C L, Tsai L Y, et al. Gigantic jets with negative and positive polarity streamers. J Geophys Res, 2010, 115, doi: 10.1029/2009JA014831

Chern, J. L., R. R. Hsu, H. T. Su, S. B. Mende, H. Fukunishi, Y. Takahashi, and L. C. Lee (2003), Global survey of upper atmospheric transient luminous events on the ROCSAT-2 satellite, J. Atmos. Solar Terr. Phys., 65, 647–659, doi:10.1016/S1364-6826(02)00317-6.

Cummer, S. A., U. S. Inan, T. F. Bell, and C. P. Barrington-Leigh (1998), ELF radiation produced by electrical currents in sprites, Geophys. Res. Lett., 25, 1281.

Cummer, S. A., 2003, Current moment in sprite-producing lightning. J. Atmos. Solar. Terr. Phys., 65, 499-508.

Cummer S A, Li J, Han F, et al. Quantification of the troposphere-to-ionosphere charge transfer in a gigantic jet. Nature Geosci, 2009, 2: 617-620, doi: 10.1038/NGEO607

Cummins, K. L., Murphy, M. J., Bardo, E. A., Hiscox, W. L.,Pyle, R. B., 1998, A combined TOA/MDF technology upgrade of the U.S. National Lightning Detection Network. J. Geophys. Res., 103(D8), 9035-9044.

Farges, T., E. Blanc, A. L. Pichon, T. Neubert, and T. H. Allin (2005), Identification of infrasound produced by sprites during the Sprite2003 campaign, Geophys. Res. Lett., 32 (1), L01813, doi:10.1029/2004GL021212.

Feng G. L., Qie, X. S., Yuan, T., Niu, S. Z., 2007, Lightning activity and precipitation structure of hailstorms. Sci. China. Ser D-Earth. Sci., 50(4), 629-639.

Franz, R. C., Nemzek, R. J., Winckler, J. R., 1990, Television image of a large upward electrical discharge above a thunderstorm system. Science, 249(4964),

48-51.

Frey, H. U.*, S. B. Mende, S. A. Cummer, A. B. Chen, R. R. Hsu, H. T. Su, Y. S. Chang, T. Adachi, H. Fukunishi and Y. Takahashi (200507), Beta-type stepped leader of elve-producing lightning, Geophys. Res. Lett., Volume 32, Issue 13, L13824, doi: 10.1029/2005GL023080.

Fukunishi, H., Y. Takahashi, M. Kubota, K. Sakanoi, U. S. Inan, and W. A.Lyons (1996), Elves: Lightning-induced transient luminous events in the lower ionosphere, Geophys. Res. Lett., 23 (16), 2157–2160.

Fukunishi, H., Y. Takahashi, T. Adachi, R. Hsu, H. Su, A. A. Chen, S. B. Mende, H. U. Frey, and L. Lee (2004), Observations of Sprites and Elves with the ISUAL Array Photometer, AGU Fall Meeting Abstracts, 51, 04.

Fullekrug, M., and S. C. Reising (1998), Excitation of Earth-ionosphere cavity resonances by sprite-associated lightning flashes, Geophys. Res. Lett., 25, 4145–4148.

Ganot, M., Yair, Y., Price, C., Ziv, B., Sherez, Y., Greenberg, E., Devir, A., Yaniv, R., 2007, First detection of transient luminous events associated with winter thunderstorms in the eastern Mediterranean. Geophy. Res. Lett., 34 (L12801), doi:10.1029/2007GL029258.

Gerken E, Inan U, Barrington-Leigh C (2000) Telescopic imaging of sprites. Geophys Res Lett 27:2637–2640. doi:10.1029/2000GL000035

Gerken, E. A., and U. S. Inan (2002), A survey of streamer and diffuse glow dynamics observed in sprites using telescopic imagery, J. Geophys. Res., 107 (A11), 1344, doi:10.1029/2002JA009248.

Gerken, E. A., and U. S. Inan (2003), Observations of decameter-scale morphologies in sprites, J. Atmos. Solar Terr. Phys., 65, 567–572, doi:10.1016/S1364-6826(02)00333-4.

Greenberg, E., Price, C., Yair, Y., Ganot, M., Bor, J., Satori, G., 2007, ELF transients associated with sprites and elves in eastern Mediterranean winter thunderstorms. J. Atmos. Solar. Terr. Phys., 69, 1569-1586.

Gurevich, A. V., J. A. Valdivia, G.M.Milikh., K. Papadopoulos, Runaway electrons in the atmosphere in the presence of a magnetic field, Radio. Sci., 31(1996), 1541-1554.

Hardman, S., Dowden, R. L., Brundell, J. B., Bahr, J. L., Kawasaki, Z., Rodger, C. J., 2000, Sprite observations in the northern territory of Austrarlia. J. Geophys. Res., 105 (D4), 4689-4697.

Hayakawa, M., Nakamura, T., Hobara, Y., Williams, E., 2004, Observation of sprites over the Sea of Japan and conditions for lightning-induced sprites in winter. J. Geophys. Res., 109 (A01312), doi:10.1029/2003JA009905.

Inan, U.S., T.F. Bell, and J.V. Rodriguez, Heating and ionization of the lower ionosphere by lightning, Geophys. Res. Lett., 1991, 18(4):705-708.

Inan, U.S., W.A. Sampson, and Y.N. Taranenko, Space-time structure of lower ionospheric optical flashes and ionization changes produced by lightning EMP, Geophys. Res. Lett., 1996, 23:133-136.

Inan, U. S., C. Barrington-Leigh, S. Hansen, V. S. Glukhov, T. F. Bell, and R. Rairden (1997), Rapid lateral expansion of optical luminosity in lightning-induced ionospheric flashes referred to as 'elves', Geophys. Res. Lett., 24 (5), 583–586.

Israelevich, P. L., Y. Yair., A. D. Devir, J. H. Joseph, Z. Levin, I. Mayo, M. Moalem, C. Price, B. Ziv, and A. Sternlieb (2004), Transient airglow enhancement observed from the space shuttle Columbia during the MEIDEX sprite campaign, Geophys. Res. Lett., 31, 06124, doi:10.1029/2003GL019110.

Krehbiel, P. R., J. A. Riousset, V. P. Pasko, R. J. Thomas, W. Rison, M. A. Stanley, and H. E. Edens (2008), Upward electrical discharges from thunderstorms, Nat. Geosci., 1, 233–237, doi:10.1038/ngeo162.

Kummerow, C., Barnes, W., Kozu, T., Shiue, J., Simpson, J., 1998, The Tropical Rainfall Measuring Mission (TRMM) Sensor Package. J. Atmos. Oceanic. Technol., 15(3), 809-817.

Kuo, C.L., R. R. Hsu, A. B. Chen, et al. Electric fields and electron energies inferred from the ISUAL recorded sprites, Geophy. Res. Lett., 2005, 32(L19103), doi:10.1029/2005GL023389.

Kuo, C. L.*, A. B. Chen, Y. J. Lee, L. Y. Tsai, R. K. Chou, R. R. Hsu, H. T. Su, L. C. Lee, S. A. Cummer, H. U. Frey, S. B. Mende, Y. Takahashi and H. Fukunishi, 2007, Modeling elves observed by FORMOSAT-2 satellite, Journal of Geophysical Research: Space Physics, Volume 112, Issue A11, A11312.

Kuo,C.L., J. K. Chou., L. Y. Tsai., et al. Discharge processes, electric field, and electron energy in ISUALrecorded gigantic jets, J. Geophys. Res., 2009, 114(A04314), doi: 10.1029/2008JA013791.

Kuo, C. L.*, T. Y. Huang, S. C. Chang, J. K. Chou, L. J. Lee, Y. J. Wu, A. B. Chen, H. T. Su, R. R. Hsu, H. U. Frey, S. B. Mende, Y. Takahashi and L. C. Lee, 2012, Full-kinetic elve model simulations and their comparisons with the ISUAL observed events, J. Geophys. Res., Volume 117, A07320.

Lang, T. J., Lyons, W. A., Rutledge, S. A., Meyer, J. D., MacGorman, D. R., Cummer, S. A., 2010, Transient luminous events above two mesoscale convective systems: Storm structure and evolution. J. Geophys. Res., 115(A00E22), doi:10.1029/2009JA014500.

Liu, D. X., Qie, X. S., Xiong,Y. J., Feng, G. L., 2011, Evolution of the total lightning activity in a leading-line and trailing stratiform mesoscale convective system over Beijing. Adv. Atmos. Sci., 28(4), 866–878, doi: 10.1007/s00376-010-0001-8.

Liu, F.X., Li, X.Y., Cao, Z.X., Chen, J.X., Ma, Q.M., 1997. The application & analysis of data of lightning location system in Shandong province. Shandong Electr. Power 96 (4), 1–8 (In Chinese).

Liu Ningyu, Nicholas Spiva, Joseph R. Dwyer, Hamid K. Rassoul, Dwayne Free, Steven A. Cummer, Upward electrical discharges observed above Tropical Depression Dorian, Nature Communications, 2015, 6, 5995, doi: 10.1038/ncomms6995.

Liszka, L. (2004), On the possible infrasound generation by sprites, J. Low Freq.

Noise, Vibration and Active Cont., 23 (2), 85–93.

Lyons, W. A., 1996, Sprite observations above the U.S. High Plains in relation to their parent thunderstorm systems. J. Geophys. Res., 101(D23), 29,641-29,652.

Lyons, W. A., Williams, E. R., Cummer, S. A., Stanley, M. A., 2003, Characteristics of sprite-producing positive cloud-to-ground lightning during the19 July 2000 STEPS mesoscale convective systems. Mon.Weather. Rev., 131, 2417-2427.

Marshall, R.A., Inan, U. S., Lyons, W.A., 2007,Very low frequency sferic bursts, sprites, and their association with lightning activity. J. Geophys. Res.,112(D22105), doi:10.1029/2007JD008857.

Mende, S. B., R. L. Rairden, G. R. Swenson, and W. A. Lyons (1995), Sprite spectra: N2 1 PG band identification, Geophys. Res. Lett., 22, 2633–2637.

Mende SB, Frey HU, Rairden RL, Su H-T, Hsu R-R, Hsu R-R, Hsu R-R, Hsu R-R, Hsu R-R, Hsu R-R, Allin TH et al (2002) Fine structure of sprites and proposed global observations, Cospar Colloquia Series. In: Liu LH (ed) Space Weather Study using Multipoint Techniques, vol 12. Pergamon Elsevier Science, pp 275–282

Mende, S. B., H. U. Frey, R. R. Hsu, H. T. Su, A. B. Chen, L. C. Lee, D. D. Sentman, Y. Takahashi, and H. Fukunishi (2005), D region ionization by lightninginduced EMP, J. Geophys. Res., 110, A11312, doi:10.1029/2005JA011064.

Hampton, D. L., M. J. Heavner, E. M.Wescott, and D. D. Sentman (1996), Optical spectral characteristics of sprites, Geophys. Res. Lett., 23, 89–93.

Hsu R R, Su H T, Chen A B, et al. Transient luminous events in the vicinity of Taiwan. J Atmos Solar Terr Phys, 2003, 65(5), 561-566.

Ignaccolo M, Farges T, Blanc E, Fu"llekrug M (2008) Automated chirp detection with diffusion entropy: Application to infrasound from sprites. Chaos Solitons Fractals 38:1039–1050. doi:10.1016/j.chaos. 2007.02.011.

McHarg MG, Stenbaek-Nielsen HC, Kammae T (2007) Observation of streamer formation in sprites. Geophys Res Lett 34:L06804. doi:10.1029/2006GL027854.

Moudry, D. R., H. C. Stenbaek-Nielsen, D. D. Sentman, and E. M. Wescott (2003),

Imaging of elves, halos and sprite initiation at 1 ms time resolution, J. Atmos. Solar Terr. Phys., 65, 509–518, doi:10.1016/S1364-6826(02)00323-1.

Morrill, J., E. Bucsela, C. Siefring, M. Heavner, S. Berg, D. Moudry, S. Slinker, R. Fernsler, E. Wescott, D. Sentman, and D. Osborne (2002), Electron energy and electric field estimates in sprites derived from ionized and neutral N2 emissions, Geophys. Res. Lett., 29 (10), doi:10.1029/2001GL014018.

Neubert, T., Allin, T. H., Stenbaek-Nielsen, H., Blanc, E., 2001, Sprites over Europe. Geophys. Res. Lett., 28 (18), 3585-3588.

Papadopoulos, K., J. Valdivia, High altitude discharges and gamma-ray flashes: a manifestation of runaway breakdown-Comment, Geophys. Res. Lett., 24(1997), 2643-2644

Papadopoulos, K., G. Milikh., J. Valdivia, Can gamma radiation be produced in the electrical environment above thunderstorms-Comment, Geophys. Res. Lett., 23(1996),2283-2284.

Pasko V P, Stanley M A, Mathews J D, et al. Electrical discharge from a thunderstorm top to the lower ionosphere. Nature, 2002, 416: 152-154

Pasko, V. P., U. S. Inan, and T. F. Bell (1996), Blue jets produced by quasielectrostatic pre‐discharge thundercloud fields, Geophys. Res. Lett., 23, 301–304, doi:10.1029/96GL00149.

Pasko, V. P., and J. J. George (2002), Three‐dimensional modeling of blue jets and blue starters, J. Geophys. Res., 107(A12), 1458, doi:10.1029/2002JA009473.

Petrov, N. I., and G. N. Petrova (1999), Physical mechanisms for the development of lightning discharges between a thundercloud and the ionosphere, Tech. Phys., 44, 472–475, doi:10.1134/1.1259327.

Pinto, O. Jr., Saba, M. M. F., Pinto, I. R. C.A., Tavares, F. S. S., Naccarato, K. P., Solorzano, N. N., Taylor, M. J., Pautet, P. D., Holzworth, R. H., 2004, Thunderstorm and lightning characteristics associated with sprites in Brazil. Geophys. Res. Lett., 31(L13103), doi:10.1029/2004GL020264.

Price, C., E. Greenberg, Y. Tair, G. Satori, J. Bor, H. Fukunishi, M. Sato, P.

Israelevich, M. Moalem, A. Devir, Z. Levin, J. H. Joseph, I. Mayo, B. Ziv, and A. Sternlieb (2004), Ground-based detection of TLE-producing intense lightning during the MEIDEX mission on board the space shuttle columbia, Geophys. Res. Lett., 31, L20107, doi:10.1029/2004GL020711

Raizer, Y. P., G. M. Milikh, and M. N. Shneider (2006), On the mechanism of blue jet formation and propagation, Geophys. Res. Lett., 33, L23801, doi:10.1029/2006GL027697.

Raizer, Y. P., G. M. Milikh, and M. N. Shneider (2007), Leader streamers nature of blue jets, J. Atmos. Sol. Terr. Phys., 69, 925–938, doi:10.1016/j.jastp.2007.02.007.

Reising, S. C., U. S. Inan, and T. F. Bell (1999), ELF sferic energy as a proxy indicator for sprite occurrence, Geophys. Res. Lett., 26, 987–990.

Riousset, J. A., V. P. Pasko, P. R. Krehbiel, W. Rison, and M. A. Stanley (2010), Modeling of thundercloud screening charges: Implications for blue and gigantic jets, J. Geophys. Res., 115, A00E10, doi:10.1029/2009JA014286.

Sâo Sabbas, F. T., Sentman, D. D., Wescott, E. M., O. P. Jr., O. M. Jr., Taylor, M. J., 2003, Statistical analysis of space-time relationships between sprites and lightning. J. Atmos. Solar. Terr. Phys., 65, 525-535.

Sâo Sabbas, F. T., Taylor, M. J., Pautet, P. D., Bailey, M., Cummer, S., Azambuja, R. R., Santiago, J. P. C., Thomas, J. N., Pinto, O., Solorzano, N. N., Schuch, N. J., Freitas, S. R., Ferreira, N. J., Conforte, J. C., 2010, Observations of prolific transient luminous event production above a mesoscale convective system in Argentina during the Sprite2006 Campaign in Brazil. J. Geophys. Res., 115 (A00E58), doi:10.1029/2009JA014857.

Sentman, D. D., E. M. Wescott, D. L. Osborne, et al. Preliminary results from Sprites94 campaign: 1. Red sprites. Geophys Res Lett, 1995, 22(10),1205-1208.

Soula, S., Van der Velde, O. A., Montanya, J., Neubert, T., Chanrion, O., Ganot, M., 2009, Analysis of thunderstorm and lightning activity associated with sprites observed during the EuroSprite campaigns: Two case studies. Atmos. Res., 91,

514-528.

Soula, S., Van der Velde, O. A., Palmiéri, J., Chanrion, O., Neubert,T., Montanyà, J., Gangneron, F., Meyerfeld, Y., Lefeuvre, F., Lointier, G., 2010, Characteristics and conditions of production of transient luminous events observed over a maritime storm. J. Geophys. Res., 115 (D16118), doi:10.1029/2009JD012066.

Soula S, Van der Velde O A, Montanya J, et al. Gigantic jets produced by an isolated tropical thunderstorm near Réunion Island. J Geophys Res, 2011, 116, doi:10.1029/2010JD015581

Stanley, M., P. Krehbiel, M. Brook, C. Moore, W. Rison, and B. Abrahams (1999), High speed video of initial sprite development, Geophys. Res. Lett., 26, 3201–3204.

Stanley, M., M. Brook, P. Krehbiel, and S. A. Cummer (2000), Detection of daytime sprites via a unique sprite ELF signature, Geophys. Res. Lett., 27, 871–874.

Su, H. T., Hsu, R. R., Chen, A. B., Lee, Y. J., Lee, L. C., 2002, Observation of sprites over the Asian continent and over oceans around Taiwan. Geophy. Res. Lett., 29 (4), doi:10.1029/2001GL013737.

Su H T, Hsu R R, Chen A B, et al. Gigantic jets between a thundercloud and the ionosphere. Nature, 2003, 423: 974-976

Takahashi, Y., M. Fujito, Y. Watanabe, H. Fukunishi, and W. A. Lyons (2000), Temporal and spatial variations in the intensity ratio of N2 1st and 2nd positive bands in SPRITES, Middle Atmosphere and Lower Thermosphere Electrodynamics Advances in Space Research, 26 (8), 1205–1208.

Suzuki, T., Y. Matsudo, T. Asano, M. Hayakawa, and K. Michimoto, Meteorological and electrical aspects of several winter thunderstorms with sprites in the Hokuriku area of Japan, J. Geophys. Res., 2011, 116, D06205, doi:10.1029/2009JD013358.

Suzuki, T., Hayakawa, M., Matsudo, Y., Michimoto, K., 2006, How do winter thundercloud systems generate sprite-inducing lightning in the Hokuriku area of Japan? Geophys. Res. Lett., 33(L10806), doi:10.1029/2005GL025433.

van der Velde OA, Mika A ′, Soula S, Haldoupis C, Neubert T, Inan US (2006) Observations of the relationship between sprite morphology and in-cloud lightning processes. J Geophys Res 111:D15203. doi:10.1029/2005JD006879

Van der Velde O A, Lyons W A, Nelson T E, et al. Analysis of the first gigantic jet recorded over continental North America. J Geophys Res, 2007, 112, doi: 10.1029/2007JD008575

Van der Velde, O. A., Montanyà, J., Soula, S., Pineda, N., Bech, J., 2010, Spatial and temporal evolution of horizontally extensive lightning discharges associated with sprite-producing positive cloud-to-ground flashes in northeastern Spain. J. Geophys. Res., 115(A00E56), doi:10.1029/2009JA014773.

van der Velde, O. A., J. Bór, J. Li, S. A. Cummer, E. Arnone, F. Zanotti, M. Füllekrug, C. Haldoupis, S. NaitAmor, and T. Farges (2010), Multi - instrumental observations of a positive gigantic jet produced by a winter thunderstorm in Europe, J. Geophys. Res., 115, D24301, doi:10.1029/2010JD014442

Williams, E., Downes, E., Boldi, R., Lyons, W., Heckman, S., 2007, Polarity asymmetry of sprite-producing lightning: A paradox? Radio. Sci., 42 (RS2S17), doi:10.1029/2006RS003488.

Winckler, J. R., 1995, Further observations of cloud-ionosphere electrical discharges above thunderstorms. J. Geophys. Res., 100(D7), 14,335-14,345.

Winckler, J. R., Lyons, W. A., Nelson, T. E., Nemzek, R. J., 1996, New high-resolution ground-based studies of sprites. J. Geophys. Res., 101(D3), 6997-7004.

Winckler, J, R., 1998, Optical and VLF radio observations of sprites over a frontal storm viewed from O'Brien Observatory of the University of Minnesota. J. Atmos. Solar. Terr. Phys.,60(7-9), 679-688.

Wilson C T R, The electric field of a thunderstorm and some of its effects. 1925a, Proc.Roy.Soc.D37: 32-37.

Wilson C T R, The acceleration of beta-particles in strong electric fields such as those thundeclouds. Proc.Cambridge Phil.Soc. 1925b, 22:534-538.

Wescott, E.M., D.D. Sentman, D. Osborne, et al. Preliminary results from the Sprites 94 aircraft campaign: Blue jets. Geophys. Res. Lett., 1995, 22(10), 1209-1212.

Wescott, E.M., D.D. Sentman, M J Heavner, et al. Blue starter: brief upward discharges from an intense Arkansas thunderstorm, Geophys. Res. Lett., 1996, 23(16), 2153-2156.

Wescott, E.M., D.D. Sentman, M J Heavner, et al. Blue jets: their relationship to lightning and very large hailfall, and physical mechanism for their production, J. Atmos. Sol. Terr. Phys., 1998, 60(7-9), 713-724.

Wescott EM, Sentman D, Heavner M, Hampton D, Lyons WA, Nelson T (1998) Observations of 'Columniform'

sprites. J Atmos Terr Phys 60:733. doi:10.1016/S1364-6826(98)00029-7.

Wescott, E.M., D.D. Sentman, H C Stenbaek-Nielsen, et al. New evidence for the brightness and ionization of blue starters and blue jets, *J. Geophys. Res.*, 2001,106(A10):21549-21554.

Wescott, E. M., Stenbaek-Nielsen, H. C., Sentman, D. D., Heavner, M. J., Moudry, D. R., Sâo Sabbasa, F. T., 2001, Triangulation of sprites, associated halos and their possible relation to causative lightning and micrometeors. J.Geophys.Res., 106 (A6), 10,467-10,477.

Yair, Y., Israelevich, P., Devir, A., Moalem, M., Price, C., Joseph, J., Levin, Z., Ziv, B., Sternlieb, A., Teller, A., 2004, New observations of sprites from the space shuttle. J. Geophys. Res., 109(D15201), doi: 10.1029/2003JD004497.

Yair, Y., Ganot, M., Price, C., Ganot, M., Greenberg, E., Yaniv, R., Ziv, B., Sherez, Y., Devir, A., Bór, J., Sátori, G., 2009. Optical observations of transient luminous events associated with winter thunderstorms near the coast of Israel. Atmos. Res., 91(2-4), 529-537.

Yang, J., Qie, X. S., Zhang, G. S., Zhao, Y., Zhang, T., 2008, Red sprites over thunderstorms in the coast of Shandong province, China. Chin. Sci. Bull., 53(7), 1079-1086.

Yang, J., Qie, X. S., Feng, G. L., 2013, Characteristics of one sprite-producing summer thunderstorm. Atmos. res., 2013, 127, 90-115.

I want morebooks!

Buy your books fast and straightforward online - at one of the world's fastest growing online book stores! Environmentally sound due to Print-on-Demand technologies.

Buy your books online at
www.get-morebooks.com

Kaufen Sie Ihre Bücher schnell und unkompliziert online – auf einer der am schnellsten wachsenden Buchhandelsplattformen weltweit!
Dank Print-On-Demand umwelt- und ressourcenschonend produziert.

Bücher schneller online kaufen
www.morebooks.de

OmniScriptum Marketing DEU GmbH
Heinrich-Böcking-Str. 6-8
D - 66121 Saarbrücken
Telefax: +49 681 93 81 567-9

info@omniscriptum.com
www.omniscriptum.com

www.ingramcontent.com/pod-product-compliance
Lightning Source LLC
Chambersburg PA
CBHW020449220526
45464CB00002B/929